ONE GIANT LEAP

THE EXTRAORDINARY STORY OF THE MOON LANDINGS

THIS IS A CARLTON BOOK

Text and design copyright © Carlton Books Ltd, 1998

This edition published by Carlton Books Limited, 1998

A CIP catalogue record for this book is available from the British Library

ISBN 1 85868 605 9

Executive Editor: Sarah Larter
Project Editor: Camilla MacWhannell
Project art direction: Diane Spender
Designer: Steve Wilson
Production: Sarah Schuman
Picture Research: Lorna Ainger

Printed and bound in Italy

ONE GIANT LEAP

THE EXTRAORDINARY STORY OF THE MOON LANDINGS

TIM FURNISS

CONTENTS

CHAPTER ONE

Shock Start to the
SPACE
AGE

T HE SPACE AGE BEGAN ON 4 OCTOBER 1957 NOT ON THE MOON BUT IN A MAN MADE CRATER, IN THE SOVIET UNION.

BELOW *Explorer 1*, the first American satellite, was launched by a *Jupiter C* booster from pad 56 at Cape Canaveral on 31 January 1958.

The location was 402 km (250 miles) east of the Aral Sea on the barren, stark steppe land of Kazakhstan. At the end of a railway track leading from Tyuratam junction on the Moscow to Tashkent main line, an open cast mine had been converted into the flame trench beneath the launch pad for the world's first intercontinental ballistic missile (ICBM).

That short and stubby missile had been test-fired secretly in May 1957 and was followed by seven flights, ending in August, when the Soviet Union announced that it had an operational ICBM, much to the consternation of the West. The ninth ICBM did not have a dummy nuclear warhead but a silver globe weighing 83.5 kg (184 lbs). This missile was launched into the night sky and minutes later, the globe was in an orbit around the Earth, ranging between about 402 km (250 miles) and 804.5 km (500 miles). The first artificial satellite. *Sputnik 1*'s radio transmitter began to emit its now-legendary "bleep, bleep" signals which could be heard by tracking stations over the world as the satellite passed overhead, travelling at about 28,158 kmph (17,500 mph). The Space Age had begun, not with awe-struck wonderment and dreams of space stations and flights to the Moon but with a great shock because the first satellite was a Soviet one. The first was expected to have been orbited by the United States.

Although the Soviet Union had announced in 1955 – along with the USA – that it would launch a satellite, no one seriously believed that the Soviets were capable of such a feat and to do it using an ICBM, which the USA did not yet possess, caused a shock that soon turned into Cold War paranoia. Not only could the Soviets hit the USA with a nuclear missile, they had the potential to control space. The next launch did not help matters for the fearful West. On 3 November, a similar missile launched the much heavier *Sputnik 2*, which this time contained the first living creature in orbit, a canine bitch called Laika, who was to die from lack of air after about six days in

space. Not only was there a missile gap but space was being dominated by the Soviets.

AMERICA'S COMEBACK

All eyes turned to America's response, the *Vanguard* satellite, the size of a grapefruit and to be launched on a less powerful and rather slim and elegant rocket of the same name. On 6 December, on a launch pad at Cape Canaveral, a sand pit protruding into the Atlantic Ocean half way up the east coast of Florida, the *Vanguard* rocket ignited. It rose a few metres, then fell back, its engine losing thrust, and broke apart in a breathtaking explosion. America's answer to *Sputnik* had failed miserably and unlike the Soviets' secret efforts, was there to be witnessed by the whole world, live on radio and TV in the USA and in front of international agency reporters.

The pressure was on America to catch up because there was a real sense of fear about Soviet domination. A different type of rocket was used for the next American attempt. This was called the *Jupiter C*, a

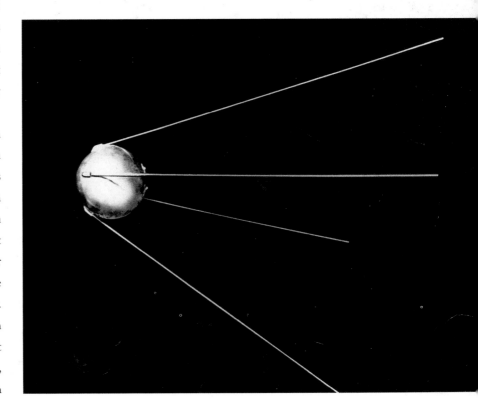

BELOW *Sputnik 1*, the world's first artificial satellite, was launched by the USSR on 4 October 1957.

8

THE FIRST SATELLITE LAUNCH ATTEMPTS

NAME	DATE	COUNTRY	RESULT
Sputnik 1	4 Oct 1957	USSR	Orbit
Sputnik 2	3 Nov 1957	USSR	Orbit
Vanguard	6 Dec 1957	USA	Failed
Explorer 1	31 Jan 1958	USA	Orbit
Vanguard	5 Feb 1958	USA	Failed
Explorer 2	5 Mar 1958	USA	Failed
Vanguard 1	17 Mar 1958	USA	Orbit
Explorer 3	26 May 1958	USA	Orbit
Sputnik	27 Apr 1958	USSR	Failed
Vanguard	28 Apr 1958	USA	Failed
Sputnik 3	15 May 1958	USSR	Orbit
Vanguard	27 May 1958	USA	Failed
Vanguard	26 Jun 1958	USA	Failed
Explorer 4	26 Jul 1958	USA	Orbit
Explorer 5	24 Aug 1958	USA	Failed
Vanguard	26 Sep 1958	USA	Failed
Beacon	23 Oct 1958	USA	Failed
Score	18 Dec 1958	USA	Orbit

modified military missile called an intermediate range ballistic missile (IRBM) which did not have the range of an ICBM. The *Jupiter C* was launched on 31 January 1958 and successfully placed the *Explorer 1* science satellite into a higher orbit than that reached by the *Sputnik* satellites and one in which the satellite made one of the most important early discoveries in space. *Explorer 1*'s Geiger counter found that the Earth was encircled by two belts of radiation. These became known as the van Allen radiation belts after the scientist who developed the science package on the *Explorer* spacecraft. A *Vanguard* satellite eventually made it into orbit in March 1958 and is today the oldest object still in space, the two *Sputniks* and *Explorer 1* having re-entered the Earth's atmosphere in 1958 and 1970 respectively. The Soviets followed with *Sputnik 3* in May 1958, carrying a suite of science experiments. America also successfully launched further satellites.

RIGHT Laika the dog was the first living creature in orbit, carried there by Sputnik 2.

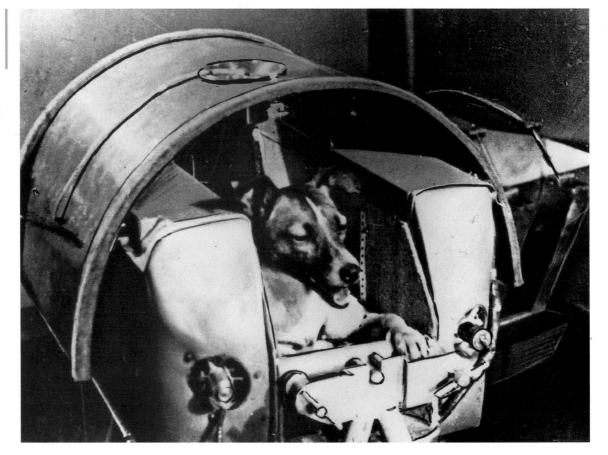

The American and Soviet spacecraft proved that space exploration was beneficial and later many other different types of satellites were launched to study the world's weather, provide international communications and navigation services, applications of space that were to be overshadowed by what would become known as the Space Race. Comparisons were being made in 1958 between the number of successful satellites being launched by the Soviet Union and the number of failed launches by America, creating the impression of Soviet dominance in space. The Soviets tried four launches between 1957 and 1958 and failed once, while the USA tried 14 times with nine failures.

The scoreboard was actually: Soviets three. USA five. The Soviet failure, however, was not revealed until many years later.

Politicians began to take note of how important space was becoming as a factor in international relations and the seeds were being sown for more spectacular events in space. Scientists and space dreamers had always regarded the Moon as the first target to reach in space and predictions were being made of space colonies on the lunar surface. The Space Age was still in its infancy as unmanned space probes were prepared for their first flights in 1958, initiating a new challenge – the Race to the Moon.

ABOVE *Vanguard 1*, launched in March 1958, is still in orbit, the oldest satellite in space.

CHAPTER TWO

The Tantalizing
TARGET

THE MOON HAS STIRRED HUMAN EMOTIONS SINCE THE FIRST PEOPLE WATCHED ITS REGULAR, RELIABLE APPEARANCES IN THE NIGHT SKY.

It has been a part of everyone's life through the ages. So close and familiar but untouchable. Twentieth-century dreamers had no doubt that humans would fly to the Moon one day, perhaps to establish a Moon base, another place to live in the cosmos.

The launch of the first satellites spurred rocket scientists into action and with only five satellites safely in orbit, the first attempt was made to send a spacecraft from the Earth to our nearest neighbour in space. What was the Moon really like? Could man land there one day?

BELOW *Pioneer 2* was launched in November 1958 but only reached 1550 km (968 miles).

USA MOON PROBES: PIONEER

The objective of America's *Pioneer 1* was highly ambitious – not just hit the Moon but to orbit it! It's *Thor Able* booster thundered into Cape Canveral's skies on 17 August 1958 but 77 seconds later exploded in a massive ball of flame. The Soviets suffered several similar failures but no one in the West was to know at the time. America's *Pioneer 1B* was launched in October and travelled 300,000 km (184,500 miles) towards its target. It did not have the velocity needed to fully escape the pull of Earth's gravity and was dragged back to burn up in the atmosphere. The craft returned some vital data from deep space about the radiation belts that had been discovered by *Explorer 1*.

Pioneer 2 reached only 1,558 km (968 miles) after its launch in November but a smaller *Pioneer* was launched in December, by a *Juno 2*, on a mission to fly past the Moon and make some scientific measurements en route. However, the Moon was proving to be an elusive target. *Pioneer 3* reached about 103,000 km

THE FIRST ATTEMPTS TO REACH THE MOON

NAME	DATE	COUNTRY	RESULT
Pioneer 1A	17 Aug 1958	USA	Failed
Luna	23 Sep 1958	USSR	Failed
Pioneer 1B	11 Oct 1958	USA	Failed
Luna	12 Oct 1958	USSR	Failed
Pioneer 2	8 Nov 1958	USA	Failed
Pioneer 3	6 Dec 1958	USA	Failed
Luna 1	2 Jan 1959	USSR	Missed Moon
Pioneer 4	3 Mar 1959	USA	Success. Lunar fly-by
Luna	18 Jun 1959	USSR	Failed
Luna 2	12 Sep 1959	USSR	Success. Hit Moon
Pioneer	24 Sep 1959	USA	Failed
Luna 3	4 Oct 1959	USSR	Success. Lunar loop
Pioneer	26 Nov 1959	USA	Failed
Luna	12 Apr 1960	USSR	Failed
Pioneer	25 Sep 1960	USA	Failed
Pioneer	15 Dec 1960	USA	Failed

in the West, however, was that *Luna 2* came from the East and that metallic spheres marked with Soviet emblems were scattered across the lunar dust. *Luna 2* perpetuated the "*Sputnik* effect".

None of these probes had carried cameras but a new series of craft was introduced by America and the Soviet Union to return the first clear views of the Moon from close quarters. The Soviets' *Luna 3* was the first to fly on 4 October, coincidently, the second anniversary of *Sputnik 1*. It surprised and enthralled the world. The craft entered an orbit around the Earth that also took it around the Moon during some of its orbits. This meant that *Luna 3* flew around the far side of the Moon. Its cameras took about 30 exposures and developed them on board, transmitting them to Earth.

For the first time in human history the hidden face of the Moon was revealed – at least 70% of it – showing it to be much more heavily cratered than the visible side and with fewer "seas", the smooth areas of the Moon dubbed by early astronomers as Mare or sea because they resembled oceans as seen from the Earth. America's first attempts to take images of the Moon followed. The first launch of a *Pioneer* spacecraft designated *Atlas Able 4*, after its launcher, failed in November 1959 and was followed by two more catastrophes. Thus the first phase of lunar exploration ended.

Like an excited child with a new toy, rocket scientists had completed a frenzy of launches towards the obvious first target, not just to explore the tantalizing Moon – but because it was there.

(64,000 miles) and fell back to Earth. At the fifth attempt, the Soviets managed to despatch a *Luna* probe towards the Moon on 2 January 1959.

USSR MOON PROBES: LUNA

Luna 1 was supposed to impact the Moon, scattering Soviet emblems upon its surface but missed by about 5,712 km (3,550 miles) and sailed into solar orbit to become the first artificial planet. Nonetheless, amid the frenzy of the Space Race, it was hailed as a success. America's *Pioneer 4* followed *Luna 1* into solar orbit after missing the Moon by a distance of 60,350 km (37,500 miles) in March. *Pioneer* had, however, been the most successful of the probes as it missed its fly-by target by only 24,140 km (15,000 miles) – good shooting in those early days.

Finally, on 14 September, a man-made object from the Earth reached the Moon. The Soviet Union's *Luna 2* was smashed to smithereens, during a planned impact at a speed of 3 km per second (2 miles/sec) and at a 30° angle, close to the crater Archimedes on the edge of the Mar Imbrium. The impact was hailed as the greatest achievement yet in the Space Age. *Luna 2* had demonstrated the accuracy required to fly to the Moon and had also sent back information from its magnetometer, and radiation and micrometeroid detectors before impact. What was far more significant

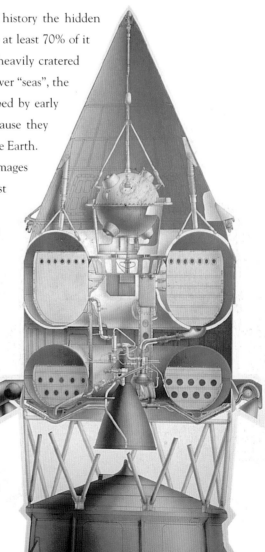

The Race
to Get
IN MAN
IN SPACE

BY 1959, IT WAS CLEAR THAT SPACE WAS A "POLITICAL BATTLEFIELD" AND AN INTEGRAL PART OF THE COLD WAR: WHOEVER DOMINATED EARTH'S ORBIT WOULD HAVE SUPREMACY.

The Soviets launched the first satellite and the first craft to hit the Moon. Who would launch the first man into space?

Before *Sputnik*, the concepts for manned space vehicles followed a logical approach. Winged space ferries flying to and from space stations featured a lot in early concepts. America was mastering rocket plane technology and had developed the piloted *X-15* winged rocket plane which was poised to reach the edge of space. The *X-15* was seen as a precursor to the first American manned spacecraft. It would take years to develop such a vehicle and that just would not do in the Space Race in which the next propaganda target was the first manned spaceflight. This objective was dubbed seriously by some American officials as "Man in Space Soonest". This meant that, quite simply, the first American space travellers would fly in recoverable capsules flying like nosecones on the top

ABOVE The *Sputnik* booster was designed to carry unmanned *Luna* and manned *Vostok* spacecraft.

LEFT Back (l to r): Shepard, Grissom, Cooper. Front: Schirra, Slayton, Glenn, Carpenter.

RIGHT Strelka and Belka were the first living creatures to be recovered from orbit.

of rockets and not in rocket planes. Secretly, the Soviets were following exactly the same rapid approach.

THE MERCURY PROGRAMME

The new American space agency called NASA – National Aeronautics and Space Administration – selected seven pilots to train for flights in the one-man capsule called *Mercury*. In the Soviet Union twelve cosmonauts began training for flights on a spacecraft that was to become known as the *Vostok*. The *Mercury* craft was far more sophisticated than the *Vostok*, a fact that only became apparent much later.

Mercury was a bell-shaped caspule that could be controlled in space by its pilot, manoeuvring in three axes called pitch, roll and yaw. The pilot could take full manual control or just monitor automatic systems. He also had the ability to override systems and troubleshoot problems. The *Mercury* had an ablative heatshield on its blunt end which would take the brunt of the intense heat on its high speed re-entry

into the Earth's atmosphere, carefully controlled at a specific angle. A parachute would enable the craft to splash down in the sea.

The first manned flights would involve suborbital "up and down" rides launched on a *Redstone* rocket and would be followed by orbital flights on the *Atlas*. The *Atlas* was America's first ICBM which had at last made its debut.

THE VOSTOK PROGRAMME

The *Vostok* was a ball-shaped capsule covered with a heatshield and attached to an instrument section. The cosmonaut would have limited control over the systems and the flight was largely controlled from the ground. The launcher would be an improved version of the *Sputnik* launcher, again based on the Soviet ICBM. The *Vostok* had to land in the Soviet Union for secrecy and political reasons and, because it would impact so hard despite its parachute, it was deemed safer to have the cosmonaut eject before it hit the ground and to land with his own parachute. Before the astronauts or cosmonauts could fly there were several test flights, many of them flown by American chimpanzees and Soviet dogs as test subjects. This only served to demonstrate that however sophisticated the craft may have seemed, they were still pretty basic machines and certainly not like the spaceships of science fiction.

Both the *Vostok* and *Mercury* test flights had mixed results, the *Vostok*s that succeeded taking most of the publicity. Three *Vostok*s failed to get into orbit before *Sputnik 4* reached orbit in May 1960. It was not actually a *Sputnik* of course and was designated No 4 to hide the fact that three launches had failed. *Sputnik 4* failed in orbit and had it been a manned flight, the cosmonaut would have been lost in space. *Sputnik 5* safely brought back two dogs, Strelka and Belka, from orbit on 20 August – the first living creatures to be recovered from orbit, although several animals had made suborbital flights before this. Two dogs were killed during a re-entry accident in December and another launch failed but two more one-dog missions were flown in March 1961. It was clear that a cosmonaut was going to be next.

On the American side, the major milestone was the *Redstone*-boosted suborbital flight of the chimpanzee Ham, in January 1961, although it was not totally successful. Because of this, it was decided to fly one more *Redstone* test in March instead of the first flight of an astronaut. America had not yet launched a *Mercury* into orbit. The excitement was reaching fever pitch, as the world waited for the first manned flight. Would it be an American or a Soviet citizen? The world found out on 12 April 1961 and the result of this race for space was to be the major influence in America's decision to shoot for the Moon

SOVIET UNMANNED VOSTOK ORBITAL TEST FLIGHTS			
NAME	DATE	COUNTRY	RESULT
Vostok	18 Jul 1959	USSR	Failed
Vostok	15 Apr 1960	USSR	Failed
Vostok	16 Apr 1960	USSR	Failed
Sputnik 4 (Vostok)	15 May 1960	USSR	Stranded in orbit
Vostok	28 Jul 1960	USSR	Failed
Sputnik 5 (Vostok)	19 Aug 1960	USSR	Success. Dogs recovered
Sputnik 6 (Vostok)	1 Dec 1960	USSR	Dogs lost in re-entry
Vostok	22 Dec 1960	USSR	Failed
Sputnik 9 (Vostok)	9 Mar 1961	USSR	Success. Dog recovered
Sputnik 10 (Vostok)	25 Mar 1961	USSR	Success. Dog recovered

Note: The USA conducted many suborbital tests but did plan orbital flights during this period

15

RIGHT Sam the rhesus monkey after a suborbital *Mercury* test flight in 1959.

CHAPTER FOUR

The FIRST SPACEMAN

ON 12 APRIL 1961 A ROCKET WAS LAUNCHED FROM THE SPUTNIK PAD AT TYURATAM.

It was similar to the one that launched *Sputnik* but had an extra stage which was originally introduced to launch the *Luna* probes. This time its payload was a *Vostok* capsule, carrying a passenger, Air Force Lt Yuri Gagarin. Once Gagarin reached orbit, the Soviets

RIGHT The first man in space: Lt Yuri Gagarin of the Soviet Air Force.

announced the momentous news. A man was in space. The world's press coverage reflected that this was one of the most important moments in history. Gagarin felt comfortable in the weightlessness of space and enjoyed the view from his orbit, which reached a maximum altitude of 327 km (203 miles). The 27-year-old pilot had no control over the craft as it made one orbit, fired a retro rocket and re-entered the Earth's atmosphere.

Before its fiery descent, the *Vostok* spun out of control almost causing disaster, a fact not revealed until over 30 years later. As the spherical capsule plunged towards the Earth before its drogue and main parachutes were deployed, Gagarin ejected as planned, leaving the craft 108 minutes into the flight but landing by parachute ten minutes later in a field in Smelovaka, near Saratov. To have the flight fully ratified as a manned mission, the Soviets had to say, wrongly, that Gagarin landed in his spacecraft and did not leave it before landing.

Gagarin had become the first hero of the Space Age and the Soviet premier Nikita Khruschev, who had drooled over *Sputnik*, was ecstatic and milked all the propaganda he could from the flight. As a further blow to American pride, the first unmanned *Mercury* capsule destined for orbit, ended up in the ocean miles offshore from Cape Canaveral when its *Atlas* booster exploded spectacularly on 25 April. An orbital flight by an astronaut seemed far off but there was still a chance of a suborbital ride into mini-history.

America's first manned flight was made on 5 May when US Navy Commander Alan B. Shepard made a brief 15-minute flight aboard the *Mercury* capsule *Freedom 7*, splashing down in the Atlantic Ocean, inside his craft. America's answer to Gagarin was no match. Shepard had not gone into orbit. The Soviets were still in the lead in space. It was obvious that space had become a theatre of international politics and one being watched by millions of people all over the world.

THE KENNEDY RESPONSE

This was not lost on a newly-elected President of the United States. Twenty days after Shepard's space hop,

ABOVE President Kennedy awarding the NASA Distinguished Service Award Medal to Astronaut John Glenn.

President John F. Kennedy responded to the Soviet lead by committing the United States to a programme to ensure absolutely that Soviet dominance in space would be beaten and that America would be regarded as the technological leader of the world.

"I believe this nation should commit itself, before this decade is out, to the goal of landing a man on the Moon and returning him safely to the Earth", he told Congress on 25 May 1961. Kennedy's goal was to be known as Project *Apollo*, the greatest technological undertaking in history. It was a crash programme costing over $25 billion over a period of eight years to prepare the way and finally to land men on the Moon before the end of 1969. The assumption that the Soviets also planned a similar exercise was correct. The Space Race was transformed into the Race to the Moon, which became an inherent part of the character of the 1960s, indeed helping to create its reputation as an exciting, fast-changing decade, nicknamed the "Swinging Sixties".

Space was never far from the newspaper headlines, as every launch and twist and turn was seen as a significant factor in the Race to the Moon. The first missions to maintain the high profile coverage were further flights of the *Vostok* and *Mercury* spacecraft, highlighted by the flight of the first American in orbit, John Glenn, who was launched by an *Atlas* booster on 20 February 1962 in a capsule he named *Friendship 7*. Glenn made three epic orbits which did not compare much to the 17 made by Soviet cosmonaut Gherman Titov in *Vostok 2* the previous August.

THE FIRST WOMAN IN SPACE

Glenn was followed by three further *Mercury* orbital missions and there were four more Soviet *Vostok* flights, one featuring the first woman in space, Valentina Tereshkova launched on 16 June 1963. Her flight was a blatant "stunt" forced on space officials by Khruschev to keep Soviet space rating high. These flights were followed avidly the world over, as the excitement of the Moon Race continued and as the designs of the spacecraft and the flight plans for Moon landing missions were being finalized both in the East and the West.

FAR LEFT *Mercury–Redstone 3* launches Alan Shepard in *Freedom 7*, 5 May 1961.

LEFT The first woman in space was the Russian Valentina Tereshkova, who flew on *Vostok 6*, launched on 16 June 1963.

19

THE FIRST MANNED SPACEFLIGHTS 1961-63					
NO.	NAME	TRAVELLER	COUNTRY	TYPE	DURATION
1	Vostok 1	Yuri Gagarin	USSR	Orbital/1	1h 48m*
2	Freedom 7	Alan Shepard	USA	Suborbital	15m 28s
3	LibertyBell 7	Gus Grissom	USA	Suborbital	15m 37s
4	Vostok 2	Gherman Titov	USSR	Orbital/17	1d 1h 18m*
5	Friendship 7	John Glenn	USA	Orbital/3	4h 55m 23s
6	Aurora 7	Scott Carpenter	USA	Orbital/3	4h 56m 5s
7	Vostok 3	Andrian Nikolyev	USSR	Orbital/64	3d 22h 25m*
8	Vostok 4	Pavel Popovich	USSR	Orbital/48	2d 22h 59m*
9	Sigma 7	Wally Schirra	USA	Orbital/6	9h 13m 11s
10	Faith 7	Gordon Cooper	USA	Orbital/22	1d 10h 19m 49s
11	Vostok 5	Valeri Bykovsky	USSR	Orbital/81	4d 23h 7m*
12	Vostok 6	Valentina Tereshkova	USSR	Orbital/48	2d 22h 50m*
*Time of ejection from spacecraft for parachute landing					

How to get to the MOON

KENNEDY'S COMMITMENT TO LAND A MAN ON THE MOON BY THE END OF 1969 WAS EXTREMELY CHALLENGING AS THE USA HAD ONLY 15 MINUTES PILOTED SPACEFLIGHT EXPERIENCE AND ONLY FIVE MINUTES OF THAT WAS ACTUALLY IN SPACE.

Kennedy was asking the USA to achieve the biggest technological feat since the completion in 1914 of the Panama Canal, which linked the Pacific and Atlantic Oceans in mid–Central America. NASA scientists, engineers and rocketeers studied the options of how to reach the Moon, considering the technical difficulties and risks, as well as the cost involved.

Three methods were looked at carefully during the rest of 1961. Over nine different Moon landing flight plans had been in the pipeline, unofficially, well before Kennedy's declaration and the Moon landing project already had been given the name, *Apollo*. Of the three "finalists", the first and the most obvious was the Direct Ascent method. This would involve the construction of a huge booster that would launch a large spacecraft and send it on a course directly to the Moon. The craft would land on the Moon and after a period of exploration, would take-off and fly directly back to the Earth.

America was already designing a huge booster called *Nova*, which would be the Moon booster. The *Nova* booster was to have been capable of generating up to 18,150,000 kg (40 million lbs) of thrust, which compared with the 166,500 kg (367,000 lbs) of the *Atlas D* that had launched John Glenn. Even if other factors had not ruled out the direct ascent method, the huge cost and technological sophistication of the *Nova* rocket quickly ruled out the option. The

FAR LEFT An early depiction of a *Saturn V* launching Apollo to the Moon

LEFT The *Apollo* capsule plunges into the Earth's atmosphere at 40,000 km/ph (25,000 mph).

Nova project was cancelled. Nonetheless, the direct ascent method was the simplest concept.

The second proposal, Earth–Orbit Rendezvous (EOR), was the logical first alternative to the Direct Ascent approach. It called for the launching of all the components required for the Moon trip into Earth orbit, where they would rendezvous, be assembled into a single system, refuelled, and sent to the Moon. This could be accomplished using a number of the smaller *Saturn C-V* launch vehicles already under development at NASA. A logical component of the EOR approach was also the establishment of a space station in Earth orbit to serve as the lunar mission's rendezvous, assembly, and refuelling point. This was an attractive option for NASA because as a result of this, a space station could emerge as the next NASA project, to serve as a scientific base and also as a jumping-off place for the exploration of the

RIGHT A 1962 depiction of what the Moon's surface would look like during an *Apollo* mission.

planets. This method of reaching the Moon, however, was also fraught with challenges, notably finding methods of manoeuvring and rendezvousing in space, assembling components in a weightless environment, and safely refuelling spacecraft.

The third option was Lunar–Orbit Rendezvous (LOR) that proposed sending the entire lunar spacecraft up in one launch. It would head to the Moon, enter into its orbit, and dispatch a small lander to the lunar surface. It was the simplest of the three methods, both in terms of development and operational costs, but it was risky. In the plan, three astronauts would be

launched in a mother ship, first reaching Earth orbit then heading for the Moon, to enter an orbit around it. A landing vehicle, manned by two astronauts, leaving one in the mother ship, would touch down on the Moon using its descent engine. After a Moonwalk, the top half of the Lunar Excursion Module (LEM) would take off, leaving the bottom half on the surface, and rendezvous and dock with the mother ship in lunar orbit. The mother ship would break out of lunar orbit and head back to Earth.

Since rendezvous was taking place in lunar, instead of Earth orbit there was no room for error or the crew

after NASA's description of "extravehicular activity", and flying in Earth orbit longer than it would take to fly to the Moon and back were also on the agenda for this next series of spacecraft which would carry two astronauts.

The project became known as *Gemini*, a constellation in the night sky represented by two major stars, Castor and Pollux who, in Greek mythology rode horses across the sky. The precedent for naming US manned spacecraft after mythological gods and heroes had been set when Project *Mercury* was established. Mercury was messenger of the gods in both Roman and Greek mythology, while Apollo was a busy fellow. He was the Greek god of archery, prophecy, poetry, music – and the Sun.

By this time, American spy satellites had provided information that indicated that Soviet missile deployments were far lower than thought at the heat of the Cold War when *Sputnik* was launched and the paranoia was perhaps unnecessary. It was also probably clear to the White House that the Russian space "lead" was not quite what it was cut out to be.

However, the impetus for the Moon landing could not stop now.

BELOW The Direct Assault method if exploring the Moon using one spacecraft was too expensive and too difficult.

could not get home. Also, some of the most difficult course corrections and manoeuvres had to be done after the spacecraft had been committed to a circumlunar flight. The LOR method was adopted in 1962 and project *Apollo* was on its way.

Rendezvous and docking of spacecraft together was to be a crucial part in the LOR method and the next series of US piloted spacecraft that succeeded *Mercury* was designed to demonstrate these manoeuvres in Earth orbit and to "rehearse" as much of a Moon flight as possible without going there. Testing out spacesuits during spacewalks, which became known as EVAs

GEMINI

24

THE GEMINI SPACECRAFT WOULD BE THE FIRST TO ALTER ITS ORBIT AND MANOEUVRE IN SPACE, WHICH WAS CRUCIAL FOR IT TO BE ABLE TO RENDEZVOUS AND DOCK.

Not surprisingly, manoeuvring was the main objective of the first piloted test flight. This was to have taken place in 1964 but was delayed until March 1965, to allow time for two crucial unpiloted tests to take place. By this time, however, the Soviet Union had sprung its usual space spectaculars to make *Gemini*'s first piloted sortie pale into insignificance. In October 1964, three cosmonauts had flown into orbit together in a spacecraft called *Voskhod 1* and just five days before *Gemini 3*, two more cosmonauts flew aboard *Voskhod 2*, one of them making the first spacewalk. America's thunder had been stolen and the Soviet Union created the impression that was already well ahead in the Moon Race. However, the *Voskhod* flights were not quite what they seemed.

ABOVE *Voskhod* was a *Vostok* capsule modified to hold up to three crew.

RIGHT Alexei Leonov was the first person to walk in space on 18 March 1965.

ABOVE *Gemini 6* is
photographed from
Gemini 7 during the first
space rendezvous in
December 1965.

LEFT Wally Schirra and
Tom Stafford are launched
in *Gemini 6* .

First, *Voskhod 1* was a one-man *Vostok* capsule crammed with three people who did not wear spacesuits. The craft had no means of emergency escape and made the most dangerous and risky spaceflight in history. It had been ordered by Premier Khruschev. If the Americans were going to launch two, then the Soviets would launch three, he said. As one of the major aims of *Gemini* was to conduct EVAs, a Soviet would just have to do it first – and did on 18 March 1965, when Alexei Leonov performed a televized spacewalk lasting about 12 minutes. Again, the craft

was a modified *Vostok* and a small, deployable airlock was built into the system. The newspapers went wild over this historic event. It was also the last "propaganda" flight, as Khruschev was overthrown in October 1964.

Gemini 3 was launched with two astronauts on 23 March for a modest three-orbit affair that successfully demonstrated the first manned manoeuvres in orbit as a crucial test for *Apollo*. Command pilot was Gus Grissom, who had flown the second suborbital *Mercury* mission in July 1961 and who was the first person to make two spaceflights. Further *Voskhod* flights were planned to match some of the *Gemini* feats, including a three-week long mission and a flight by an all-female crew but, in the first sign of Soviet weakness, these were cancelled, unbeknown publicly in the West at the time.

GEMINI'S SUCCESS

The remarkable *Gemini* programme then soared ahead with nine more piloted flights ending in November 1966, meeting all its goals during one of the the most frenetic and exciting periods of the Moon Race and one followed avidly in the press and on international TV, thanks to the communications revolution which resulted from the use of geostationary orbiting satellites, such as *Early Bird*. America's first EVA was completed by Edward White, the pilot of *Gemini 4* on 3 June 1965, during a flight with commander Jim McDivitt lasting four days. This duration was increased to eight days by *Gemini 5* in August and in December, *Gemini 7* was launched on a 14-day mission. *Gemini 7* acted as the rendezvous target for *Gemini 6* on 16 December. The two craft flew in formation as close to one foot of each other in the greatest spaceflight since Gagarin. The space foursome who achieved the feat were Frank Borman and James Lovell in *Gemini 7* and Wally Schirra – a Mercury veteran – and Tom Stafford. Astronauts Neil Armstrong and David Scott completed the first space docking on 16 March 1966 when *Gemini 8* joined up with an unmanned *Agena* target rocket, simulating the ascent of a lunar module from the Moon, docking with a mother ship in lunar orbit.

The historic achievement was marred when a thruster on *Gemini 8* short circuited and could not be turned off, spinning the craft like a catherine wheel. Armstrong and Scott were close to losing conciousness but managed to make an emergency landing. *Gemini 9*'s pilot, Gene Cernan, completed a spacewalk lasting over two hours in June but illustrated how difficult it was to work outside a spacecraft without some form of restraint to stop the tendency to float upwards. In July, *Gemini 10* used its *Agena* target vehicle's engine to boost its orbit to a record 740 km (460 miles), an altitude exceeded in a similar manner by *Gemini 11* in September, which soared to a height of 1,368 km (850 miles) over Australia.

The remarkably successful programme ended with the landing of *Gemini 12* in November and NASA felt confident that the major requirements for a Moon mission had been mastered. The Soviets had not launched a cosmonaut since March 1965. *Apollo* beckoned. Despite the assassination of President Kennedy in 1963 and America's ever-deepening involvement in the Vietnam War, the Moon momentum was still there but the Soviet Union's apparent lack of interest and a reduction in Cold War tension was becoming an issue.

GEMINI FLIGHT LOG

NO.	CREW	DATE	DURATION	RESULTS
3	Gus Grissom John Young	23 Mar 1965	4h 52m 51s	Orbit change
4	Jim McDivitt Edward White	3 Jun 1965	4d 1h 56m 12s	EVA
5	Gordon Cooper Pete Conrad	21 Aug 1965	7d 22h 55m 14s	Record duration
7	Frank Borman James Lovell	4 Dec 1965	13d 18h 35m 1s	Record duration
6	Wally Schirra Tom Stafford	15 Dec 1965	1d 1h 51m 54s	1st rendezvous
8	Neil Armstrong David Scott	16 Mar 1966	10h 41m 26s	1st docking
9A	Tom Stafford Gene Cernan	3 Jun 1966	3d 0h 20m 50s	Record EVA
10	John Young Michael Collins	18 Jul 1966	2d 22h 46m 39s	Re-boost orbit
11	Pete Conrad Dick Gordon	12 Sep 1966	2d 23h 17m 8s	Record altitude
12	James Lovell Buzz Aldrin	11 Nov 1966	3d 22h 34m 31s	Record EVA

The MOON SCOUTS

THE NEXT PHASE OF LUNAR EXPLORATION BEGAN IN JANUARY 1962 WHEN RANGER 3 WAS LAUNCHED TO LAND A CAPSULE ONTO THE SURFACE OF THE MOON.

This was the first of a series of American unpiloted scouts of various names to find out what the lunar surface was really like and whether it was a safe place

BELOW *Rangers 6–9 were launched in 1964–5 to take photographs of the Moon.*

to land. Some scientists, for example, thought that a craft would disappear in a vast layer of soft lunar dust. The Moon scouts were therefore to locate the possible landing sites. The Soviet Union also planned to continue its *Luna* series with the same objective. Unfortunately, the Moon proved to be an elusive target. *Ranger 3, 4* and *5* failed to land the capsules, although *Ranger 4* did hit the Moon's far side in April 1962.

CONTACT WITH THE MOON

A new series of *Ranger* craft was developed just to take hundreds of pictures of the Moon as the craft plunged to destruction. *Ranger 6* also failed in January 1964. By

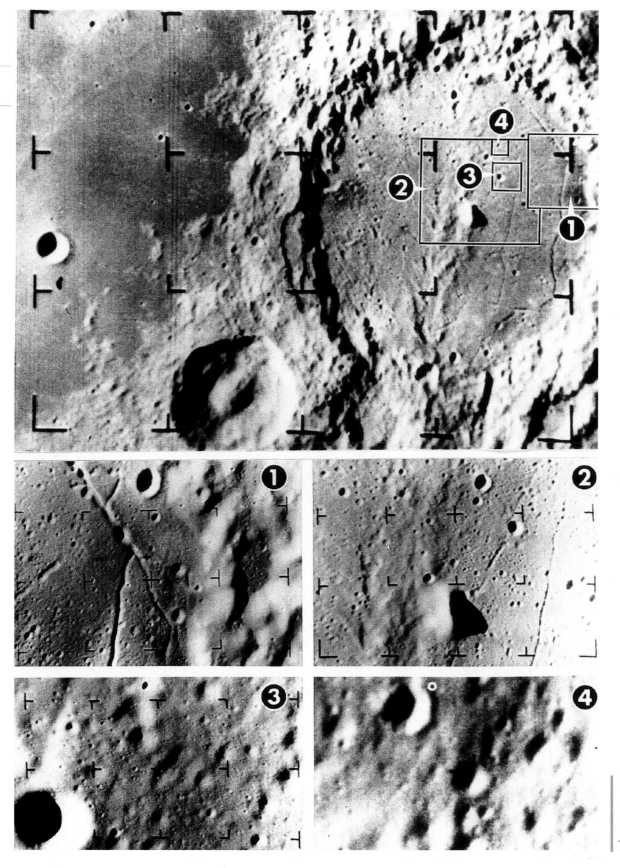

LEFT *Ranger 9* took a series of pictures of the lunar surface as it plunged into the Alphonsus crater.

this time, four Soviet *Lunas* had also failed and only one of these – *Luna 4* – was declared publicly. Finally, on 31 July 1964, *Ranger 7* smashed into the northern rim of the Sea of Clouds after returning 4,316 pictures, the last of which was taken from a few hundred feet. This was a tremendously exciting mission: for the first time mankind had a really close view of the Moon, pockmarked with thousands of craters both large and small and even boulders. It seemed as though finding a safe landing place was going to prove difficult. *Rangers 8* and 9 were also spectacular successes in 1965, with over 12,000 more images returned. *Ranger 8* viewed the Sea of Tranquillity, while *Ranger 9* was targeted at the inside of a crater, called Alphonsus. Despite the early failures, *Ranger* ended up being a smash hit.

LUNA LANDING

The Soviets launched eight more failed *Luna* probes before the epic flight of *Luna 9* which "soft-landed" on

BELOW *Surveyor 3* photographed in 1969 during the *Apollo 12* mission.

THE ELUSIVE TARGET MOON LOG 1964–1968		
NAME	**COUNTRY**	**DATE**
Succeeded		
Ranger 7	USA	28 Jul 1964
Ranger 8	USA	17 Feb 1965
Ranger 9	USA	21 Mar 1965
Luna 9	USSR	31 Jan 1966
Luna 10	USSR	31 Mar 1966
Surveyor 1	USA	30 May 1966
Lunar Orbiter 1	USA	10 Aug 1966
Luna 11	USSR	24 Aug 1966
Luna 12	USSR	22 Oct 1966
Lunar Orbiter 2	USA	6 Nov 1966
Luna 13	USSR	21 Dec 1966
Lunar Orbiter 3	USA	5 Feb 1967
Surveyor 3	USA	17 Apr 1967
Lunar Orbiter 4	USA	4 May 1967
Lunar Orbiter 5	USA	1 Aug 1967
Surveyor 5	USA	8 Sep 1967
Surveyor 6	USA	7 Nov 1967
Surveyor 7	USA	7 Jan 1968
Luna 14	USSR	7 Apr 1968
Failed		
Rangers 3, 4 and 5	USA	1962
Two Lunas	USSR	1963
Luna 4	USSR	1963
Luna	USSR	1963
Ranger 6	USA	1964
Four Lunas	USSR	1964–65
Luna 5, 6, 7 and 8	USSR	1965
Luna	USSR	1966
Surveyor 2 and 4	USA	1966–67
Luna	USSR	1968

the Moon on 3 February 1966. The craft was actually a spherical capsule ejected from the main craft just before impact. The historic *Luna 9* returned the first pictures of the surface of the Moon, showing a slightly undulating, powdery terrain, pockmarked with small craters and small rocks.

This Soviet space first was followed by the first lunar orbiter, *Luna 10*, in April 1966 and the impression of the Soviet lead in space was being perpetuated, particularly by the press.

SURVEYOR

In 1966, America started a new campaign of flights specifically supporting the *Apollo* programme, to locate and explore potential landing sites for astronauts. *Surveyor 1* made a rocket-assisted touch down in the Oceanus Procellarum on 2 June 1966, sending back spectacular mosaics of the surface. The first transmissions were covered on live TV. Dramatic advances were made in very quick succession. As *Gemini 9* was flying at the time and the first mock-up of a *Saturn 5* booster was being rolled out to the launch pad at the Kennedy Space Centre, near Cape Canaveral, America's quest for the Moon was particularly in the limelight. In 1967 *Surveyor 3* – which used a shovel to scoop up soil – and *Surveyors* 5, 6 and 7 were equally successful in other areas of the Moon, *Surveyor 7* landing close to the giant and prominent crater, Tycho.

At the same time another successful American programme, *Lunar Orbiter*, was being flown. Five consecutive successful missions proved America's technological standard, as thousands of pictures of almost the whole Moon, back and front, were sent back to the Earth between August 1966 and August 1967. *Lunar Orbiter 4*, launched on 4 May 1967, was the first to orbit the poles of the Moon, and one of its images revealed the Orientale impact basin on the far side, looking like frozen ripples on a pond. Some of the other *Lunar Orbiter* images were equally spectacular, one oblique view from an altitude of about 48 km (30 miles), showing the crater Copernicus even being described as the "Picture of the Century" by NASA. Another *Lunar Orbiter* returned the first "Earthrise" picture, of the world rising above the horizon of the Moon. The *Lunar Orbiter*s in particular enabled NASA to select up to 20 candidate landing sites for the *Apollo* programme. Meanwhile, the design and development of the rocket and spacecraft needed to get men to the Moon was nearing completion.

BELOW NASA's "Picture of the Century". *Lunar Orbiter 2*'s image showing the crater Copernicus.

The APOLLO SPACECR

AFT

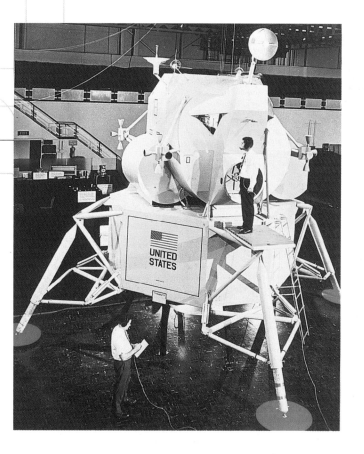

THE APOLLO LUNAR SPACECRAFT COMPRISED OF THREE MAJOR COMPONENTS, THE COMMAND MODULE, THE SERVICE MODULE AND THE LUNAR MODULE.

ABOVE An early mock up of the Lunar Module which was to land two people on the Moon.

The crew flew to and from the Moon in the Command Module; the Service Module was attached at all times to the Command Module until just before re-entry into the Earth's atmosphere, when it was jettisoned. The Lunar Module was the component in which two of the crew landed on the Moon and took-off again to join the Command Module, which remained in lunar orbit during EVAs. The Command and Service Module combination was often known as the CM and SM or the CSM and the Lunar Module, originally the LEM then just LM. Another major element of the spacecraft during the first 100 seconds of flight was the

LEFT The *Apollo* Service Module propulsion system engine blasts the spacecraft back to Earth.

KEY APOLLO SPACECRAFT TESTS

NAME	SPACECRAFT TYPE	DATE	OBJECTIVE
SA-6	Boilerplate CM	28 May 1964	Verify structure
SA-7	Boilerplate CM	18 Sep 1964	LES jettison
SA-201	CSM-009	26 Feb 1966	Re-entry
SA-202	CSM-011	25 Aug 1966	High speed entry
SA-501/Apollo	5 CSM-017	9 Nov 1967	Moon flight entry
SA-204/Apollo 4	LM-1	22 Jan 1968	Verify staging
SA-502/Apollo 6	CSM-020	4 Apr 1968	Evaluation

launch escape system in case the *Saturn* V booster malfunctioned. The system comprised a solid propellant rocket with a boost protective shield, covering the Command Module.

If an abort was ordered, the rocket would fire pulling the Command Module away from the rocket, allowing it to make a parachute landing in the sea close to the Kennedy Space Centre. If it was not required, the escape system was jettisoned after 100 seconds, taking with it the boost protective cover and

BELOW The CSM and lower adapter before joining the *Saturn* booster.

giving the crew a view from the Command Module's five windows. At the nose of the Command Module was the docking mechanism that allowed it to join up with the Lunar Module which was in fact beneath it for the launch. Once despatched towards the Moon, the CSM combination separated from the third stage of the *Saturn* V rocket, turned around and docked with the Lunar Module nestled inside the third stage, extracting it from the stage. This was called the transposition and docking manoeuvre.

INSIDE APOLLO

The combined craft flew to the Moon, with the crew able to transfer to and from the two inhabitable modules via a transfer tunnel, once the docking probe had been removed. The 5443 kg (12,000 lbs) Command Module was about 3.7 m (12 ft) high and wide, providing 7.7 cu m (235 cu ft) of comparatively luxurious space for the crew, compared with the cramped days of *Mercury* and *Gemini*. Nonetheless, it was still a small area for three crewmen, especially when they wore their bulky spacesuits during critical periods of the flight, such as launch and re-entry. The Command Module served as the flight deck, sleeping quarters, kitchen, washroom and toilet, equipped with a urine collection device which vented urine into space and the storage for solid waste bags which the crew had to use.

The display console which faced the crew from their stations on three reclining seats on the floor, measured about 2 m (7 ft) across, with switches and dials for all the systems, including the flight computer and manoeuvring thrusters and main service propulsion engine. In a small section at the foot of the couches, which also served as a private area for defecation, was the navigation bay. The cabin atmosphere was to be 15 psi, 100% oxygen. A vital part of the Command Module was the heatshield which protected the crew from 600°C (3,000°F) temperatures, experienced during the plunge into the Earth's

atmosphere, which began at a speed of about 40,225 kmph (25,000 mph). The 24,494 kg (54,000 lbs) service module was about 7.6 m (25 ft) long, with a conical rocket motor nozzle at its end and a large communications antenna on a short deployable boom. The service propulsion system engine, which was vital for the lunar orbit insertion and trans–Earth burns, had a thrust of 9,299 kg (20,500 lbs). In addition to the rocket's propellants, the service module held fuel cells, comprising liquid oxygen and liquid hydrogen which was converted into electricity, providing water as a bi-product.

The Lunar Module looked like a gangly insect, in fact it used to be called "the bug". It was a two-stage vehicle, 7 m (23 ft) high and 9.5 m (31 ft) wide across its four spindly landing legs. It weighed about 15,422 kg (34,000 lbs) but was very fragile on the Earth, comprising aluminium alloy, with a thin layer of insulation. The Lunar Module comprised the descent stage which was unmanned and contained the vital engine to perform the lunar landing. One of its legs had a ladder down which the astronauts would climb to the lunar dust. On top of the descent stage was the ascent stage, which was the flight deck for the two standing crew members, the commander on the left and lunar module pilot on the right, who both had small 3 m (1 ft) wide windows to look out of. The main hatch opened inwards and allowed the commander to exit first, crawling on his hands and knees onto a small porch on top of the ladder. Once the exploration was complete, the crew would use the descent stage as a launch pad and fire the ascent motor to fly into lunar orbit for rendezvous and docking with the CSM, manned by the lone command module pilot.

By the end of 1967, versions of the *Apollo* modules had made a number of test flights in Earth orbit using smaller *Saturn 1* and *1B* boosters. The Lunar Module was to be tested in early 1968. The first *Saturn V* had also been flight tested, proving to be an apocalyptic event.

SATURN V

ON 9 NOVEMBER 1967, THE FIRST SATURN V BOOSTER WAS LAUNCHED FROM PAD 39A AT THE KENNEDY SPACE CENTRE. IT CREATED AN EXTRAORDINARY DIN AND SENT SHOCK WAVES THROUGH THE AREA.

Five kilometres (3 miles) away at the press site, famous American TV anchor man Walter Cronkite reported as the roof of his studio began to fall in. The *Saturn V* made quite an impact. With the *Apollo* system on top, the monster rocket was 111 m (363 ft) high and its five first stage engines generated a thrust of 3,402,000 kg (7,500,000 lbs). Each engine gulped almost 13,608 kg (30,000 lbs) of propellant every second.

The rocket consisted of three stages, two of which had completed their work within nine minutes. The *SIC* first-stage *F-1* engines were powered by liquid oxygen and kerosene engines which carried the

RIGHT The *Saturn V-S4B* stage, which propelled *Apollo* out of Earth orbit.

FAR RIGHT The giant *Saturn V– Apollo* system is rolled out of the Vehicle Assembly Building.

vehicle to an altitude of 61 km (38 miles) and a speed on 13,435 kmph (8,350 mph) in 160 seconds. This was then jettisoned and the *SII* second stage was ignited. This was powered by five J2 cryogenic engines which consumed liquid oxygen and liquid hydrogen. The *SII* worked for six minutes and 30 seconds by which time the rocket was 183 km (114 miles) up and travelling at about 24,618 kmph (15,300 mph). The third stage was the re-ignitable *SIVB*, which was also powered by one cryogenic J2 engine. Depending on the flight computer, the engine burned for about 150 seconds, achieving a speed of 27,997 kmph (17,400 mph), enabling the combination to enter Earth parking orbit.

Later the J2 was restarted to set the *Apollo* on course for the Moon, on what was called the trans–lunar injection. It fired for about 300 seconds, increasing speed to 39,259 kmph (24,400 mph), or escape velocity to allow the crew to escape the pull of the Earth's gravity. After the Lunar Module had been extracted from the nose of the S4B, the spent stage sailed into deep space, sometimes entering solar orbit or even impacting on the Moon. Despite the brute force of the monster rocket, the *Saturn V* had what

FAR LEFT The night before launch. A *Saturn V* and *Apollo* await blast off for the Moon.

BELOW The interstage section is jettisoned during the launch.

THE SATURN V

Designed and developed by NASA's Marshall Space Flight Centre, Huntsville, Alabama under Dr Wernher von Braun, the former German *V2* rocket engineer

Height: 111 m (363 ft) with *Apollo* and launch escape system
Diameter: 10 m (33 ft)
Weight: 2,934,792 kg (6,470,000 lbs)
Thrust: 3,402,000 kg (7,500,000 lbs)

S-1C FIRST STAGE:

Five *F-1* engines with combined thrust of 3,402,000 kg (7,500,000 lbs), consuming 972,844 litres (214,000 gallons) of kerosene and 1,574,734 litres (346,400 gallons) of liquid oxygen, with a total weight of 2,268,000 kg (5,000,000 lbs)

S-2 SECOND STAGE:

Five *J-2* engines with a combined thrust of 453,600 kg (1,000,000 lbs), consuming 1,171,504 litres (257,700 gallons) of liquid hydrogen and 397,320 litres (87,400 gallons) of liquid oxygen, with a total weight of 74,390 kg (164,000 lbs)

S4B THIRD STAGE:

One *J-2* engine with 90,720 kg (200,000 lbs) of thrust, consuming 304,127 litres (66,900 gallons) of liquid hydrogen and and 92,738 litres (20,400 gallons) of liquid oxygen, with a total weight of 120,204 kg (265,000 lbs)

was considered to be in its day a very sophisticated flight computer. Called the instrument unit, it was a ring-like structure mounted around the rocket which measured the booster's acceleration, altitude and calculated what corrections were necessary, commanding the engines' burn time. It also measured the booster's telemetry, electrical supply and thermal conditioning system. The unit was 91 cm (3 ft) high, 6 m (21 ft) in diameter and weighed over 45 kg (100 lbs). It had the computing power of of the simplest of today's pocket calculators.

VEHICLE ASSEMBLY BUILDING AND LAUNCH PADS

To build the *Saturn Vs* and launch them required an enormous infrastructure at the Kennedy Space Centre, which is still dominated today by the building in which the *Saturn V* rockets were stacked vertically. Called the Vehicle Assembly Building or VAB, it is now used to assemble the Space Shuttle. The VAB is 160 m (525 ft) high and could fit four United Nations buildings inside. The *Saturns* were erected on Mobile Launch Platforms inside the VAB, which then moved slowly out of VAB's huge doors, on a crawler-way to either of the two launch pads, 39A or 39B. The

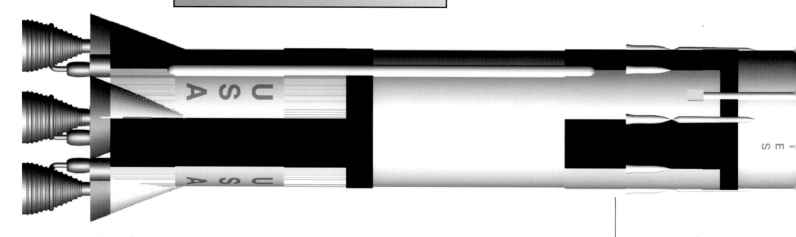

FIRST STAGE   SECOND

platforms were the "launch pads" and included 125 m (410 ft) high mobile service structure gantry towers, which had two lifts and four access platforms to prepare the rocket and spacecraft. The entire structures, weighing over 7,711,200 kg (17,000,000 lbs) were rolled down the 5.6 km (3.5 mile) crawlerways on a crawler transporter, a double-tracked vehicle the size of a football field. Each of the cleats of the tracks or "wheels" of the crawler weighs over 1,016 kg (1 ton). The system is used today for the Space Shuttle which is launched from the same two pads. The crawler transporter travels at 1.6 kmph (1 mph) has its own 15 man crew, including the driver and is powered by four large diesel engines. Once the transporter approaches the launch pad, it has to climb up a slope to place the mobile launch platform over the huge flame pit on the concrete pad. In order to keep the launch vehicle absolutely vertical on the 5° slope, the crawler is kept level by huge hydraulically operated rams.

Adjoining the VAB was the Launch Control Centre with four firing rooms each with 470 sets of consoles and monitors. The intense efforts being made to get the whole *Apollo* system in gear for the first piloted flights was illustrated by the development of the first *Saturn V* to its first launch: it took just five years. While all eyes had been focused on the very public development of the *Saturn V* and *Apollo*, the Soviet Union was finalizing its own plans for the Moon, which involved something even bigger than *Saturn V*.

LEFT A *Saturn V* is launched from the Kennedy Space Centre en route for the Moon.

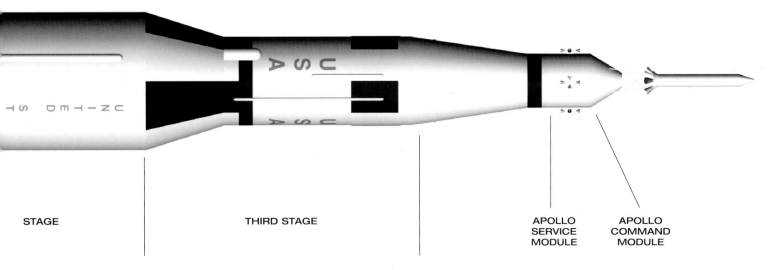

STAGE   THIRD STAGE   APOLLO SERVICE MODULE   APOLLO COMMAND MODULE

CHAPTER TEN

Soviet MOON Plans

DURING 1966, AMERICA RULED THE SPACE WAVES. THE SOVIET UNION, IT SEEMED, HAD DROPPED OUT OF PILOTED FLIGHT.

ABOVE This Soviet Lunar Module was planned to land a man on the Moon.

There had not been a cosmonaut in space since March 1965. What was happening? The Soviet Union had not pulled out. It fully intended to beat America to the Moon – in two ways. First, the Soviets would launch two cosmonauts on a round-the-Moon trip to steal American thunder; then one Soviet cosmonaut would land on the Moon by 1968, performing the *coup de gras*. To do so meant developing two new vehicles, a *Soyuz* piloted transporter and a giant rocket, called

the *N1*. The schedule was very tight but the Soviets pressed ahead with urgency.

The *Soyuz* was to be more sophisticated than the *Vostok/Voskhod*. It would be capable of carrying two or three cosmonauts and to rendezvous and dock in space. It was also to be capable of making trips to and from the Moon. It consisted of an orbital module with the docking system, a flight cabin and re-entry vehicle, and a service module, equipped with solar panels to provide on-board electrical power and its own engine for manoeuvring. The *Soyuz* would be launched by an updated version of the *Vostok* booster, also called the *Soyuz*. The *Soyuz* would be tested first in Earth orbit and conduct a docking with another *Soyuz*, eventually allowing cosmonauts to transfer externally from one to the other. The *Soyuz* without an orbital module was to be launched as a *Zond* spacecraft on figure-round-trips to the Moon and back. The *Zond* would fly unpiloted missions first, to be followed by piloted craft with two cosmonauts.

The craft would be launched on a relatively new *Proton* booster developed for unpiloted space programmes. This programme was called the *L-1*. The first prototype *Soyuz* was tested in Earth orbit unpiloted in November 1966. The lunar landing programme was called the *L-3* and was extremely ambitious. It would be based on *Soyuz/Zond* technology – and the giant *N1*. The Soviet mega booster was to be an extraordinary, almost uniformally tapered vehicle with a very long payload shroud in which the manned moon landing craft was placed. The *N1* had a base diameter of about 15 m (50 ft) and rose to a height of just over 91 m (300 ft). The 24 m (80 ft) long first stage was equipped with 30 liquid oxygen-kerosene *NK-33* engines with a thrust of about 5,080,000 kg (5,000 tons). These would operate for 110 seconds before the second stage took over. This stage was powered by eight similar engines, called the *NK-34* which were to operate for 130 seconds. The third and final stage of the booster itself was powered by similar engines called *NK-39* and which fired for 400 seconds.

The Moon landing spacecraft consisted of four main units: two more rocket motors, a lunar lander and a lunar orbiter, in which a two-person crew would fly to the Moon. The first rocket motor, called a Block G, would send the combination on a flight towards the Moon and then be discarded. The next motor, called the Block D, was to fire to place the Moon craft into lunar orbit with an eventual low point of 16 km (10 miles). Then the lunar lander cosmonaut would put on a spacesuit and make a spacewalk to transfer into the lunar module below, leaving his companion in the mother ship, which would then separate from the Block D/lander combination.

The Block D engine would fire for the descent burn to the surface but just before impact at an altitude of about 2 km (1.5 miles) above the surface, the 4.6 m (15 ft) high lander would separate and make a soft landing using its own engine, while the Block D crashed nearby. The lone cosmonuat would exit via a circular hatch and down a ladder onto the lunar surface, planting the Hammer and Sickle flag – hopefully, the first to fly on Earth's nearest neighbour. He would stay just an hour before returning to the cabin with some lunar rocks and samples and having placed some instruments on the surface. The entire lunar module would take-off from the Moon using the same engine. The lunar module would dock with the lunar orbiter and the Moonman would transfer back during an EVA. The lunar lander would be jettisoned and the lunar orbiter, based largely on a *Soyuz* capsule, would fire its engine and head for home, in a similar profile to that used by the *L-1* missions. These plans depended on successful early testing of the *Soyuz* spacecraft and later the *N1* booster.

Both US and Soviet lunar plans had reached similar stages with the US ahead in terms of rendezvous and docking and other flight experience. 1967 was to be an important year for both – for the wrong reasons.

ABOVE The Soviet Union's Mighty *N1* booster was 90 m (300 ft) high and used 30 first-stage engines.

THE CHIEF DESIGNER

The name of the Soviet designer credited with the development of the *R-7* intercontinental ballistic missile, *Sputnik*, *Vostok*, the *L-1* and *L-3* lunar flight plans and the *NI* super-booster was never revealed until his death in 1966. Previously he was just referred to as The Chief Designer. Sergei Korolev was the greatest influence in the Soviet space programme and it is fair to conclude that *Sputnik 1* would never have been launched before the USA, nor would Gagarin have beaten Shepard into space, without the force and organization of this rocket engineer. Korolev was born in the Ukraine and in 1933, after graduating from technical college, he became the founder member of a group of Soviet rocket engineers. In 1945 he led a team developing an advanced version of the *V2* captured from Germany after the Second World War. His *R-7* missile is the basis of several of the boosters being used today by Russia. Korolev also built up a skilled spacecraft design team and launched all the *Vostok* and *Voskhod* spacecraft. On 18 January 1966 he died during an operation and only afterwards was his named released to the public and his remains interred in the Kremlin Wall.

TRAGEDY
Strikes

THE GEMINI PROGRAMME WAS THE HIGHLIGHT OF A TRIUMPHANT **1966** FOR **A**MERICA.

Bold plans were afoot for 1967, including piloted test flights of the major components of the *Apollo* system, starting with a 14-day test flight of the combined CSM in Earth orbit, launched aboard a *Saturn 1B* booster, leading to a man on the Moon, possibly as

early as 1968. After a quiet 1966, the Russians were ready to start testing of the new *Soyuz* spacecraft which would provide the means to make a manned lunar-looping flight and if other components worked, including the *N1* booster, a piloted landing on the Moon. Hopes were high but behind the scenes, tragedies were in the making.

THE APOLLO 1 DISASTER

The *Apollo* CSM had been successfully tested on several types of unmanned test flights, including an important re-entry and splashdown test in 1966. *Apollo 1* was to perform a shakedown of the vehicle with a three-man crew on 14 February 1967. Commander Gus Grissom, about to make his third flight and to become the only person to fly *Mercury*, *Gemini* and *Apollo* missions, was the NASA favourite to command the first Moon mission and to become the first man on the Moon. The senior pilot was Edward White, the first American spacewalker whose

LEFT The burned-out *Apollo 1* Command Module on Pad 34 at Cape Kennedy.

FAR LEFT The *Apollo 1* crew: (l to r) Edward White, Gus Grissom and Roger Chaffee.

derring-do had captured the public's imagination in 1965. The pilot was newcomer Roger Chaffee. However, *Apollo 1* was not in great shape. Indeed the astronauts were resigned to the fact that they would hit various problems in orbit but they were test pilots and that is what test pilots do. Outwardly, however, there was an air of confidence around. It was in fact over-confidence.

On 27 January 1967, the three astronauts entered *Apollo 1* on top of its *Saturn 1B* booster on launch pad 34 at Cape Canaveral for an arduous countdown demonstration test during which the rocket was unfuelled.

Almost immediately, the astronauts encountered some expected problems, especially with the communications system. The Command Module – like *Mercury* and *Gemini* – was pressurized with pure oxygen. Bad workmanship had resulted in some electrical wiring losing its insulation and this caused a spark under Grissom's seat. Within seconds, the arc had become an inferno in the oxygen atmosphere, fed aggressively by much flammable material. White struggled to open the hatch – which in normal circumstances took 90 seconds to open – while Grissom tried to assist. With fire spreading about him, Chaffee remained in his seat monitoring systems, shouting "Fire in the spacecraft! ... we're burning up!" Poisonous carbon monoxide and other fumes entered their breathing systems and the astronauts were asphyxiated, within 15 seconds of the fire starting. The fire was so intense it had burst the bulkhead of the craft.

The *Apollo 1* disaster revealed carelessness and bad workmanship in design and production. The programme was delayed, while modifications were made, including a new quick-opening hatch and a new oxygen–nitrogen atmosphere. It looked as though America had lost the chance of landing on the moon within Kennedy's deadline. Russia had a chance to catch up. In doing so, it also planned a flight which was unnecessarily risky and in fact downright dangerous. The *Soyuz* had not even been completely and successfully tested unpiloted when it was entrusted to a cosmonaut on 23 April 1967.

THE SOYUZ 1 DISASTER

Vladimir Komarov's *Soyuz 1* flight was not to be an ordinary orbital shakedown, either; he was to dock with another piloted *Soyuz*, launched the day after. In addition, two cosmonauts from *Soyuz 2* were to space-walk to *Soyuz 1*, simulating the transfer from the Lunar Module to Lunar Lander and back, planned for the Russian Moon landing flight. The *Soyuz* extravaganza was forced on space officials by the Government. Komarov was resigned to failure and was utterly depressed when he went to the launch pad. As expected, when he reached orbit, his *Soyuz* malfunctioned. A solar panel failed to deploy, starving the craft of electrical energy and many systems also malfunctioned, including the altitude control system. Komarov was in a spin and basically helpless.

Soyuz 2 was cancelled as efforts were made to get Komarov home. Desperate attempts were made by the heroic cosmonaut to stabilize his craft for retrofire and re-entry and on orbit 18, he succeeded, resigned to perishing. He spoke with his wife in a heart-rending final conversation. The *Soyuz* re-entry flight capsule was probably damaged during a relatively uncontrolled re-entry which may have damaged the parachute deployment system. The Soviets say that the single *Soyuz* parachute tangled. The capsule, which had survived the heat of re-entry, plummeted to the steppe land. It was a dreadful end to a fated spaceflight. Russia was grounded too. The phoenix rose from the ashes for both countries in October 1968 when both *Apollo* and *Soyuz* were in orbit.

ABOVE Soyuz 1 should have looked like this in space but one solar panel failed to deploy.

"FIRE IN THE COCKPIT!"

Just after 18:31 on Friday, 27 January 1967, on Pad 34 at Cape Canaveral, a voice called from the inside of *Apollo*, "Fire! We've got fire in the cockpit!" It was Roger Chaffee. Within a second, the awful effects of a fire in a pressurized pure oxygen atmosphere took hold..... from under Gus Grissom's seat, where the first spark occurred, flames licked to the top of the cabin, surging behind the instrument panel. Netting placed in the craft rained tongues of hot, white flames down on the crew and equipment. The fire was still concentrated on Grissom's side. Ed White grabbed the handle to try to undo the six bolts that kept it well and truly shut. Alarm lights on the instrument panel lit up and a siren alarm sounded. Within seconds, the pressure and temperature had risen dramatically. "We've got a bad fire, we're burning up!" shouted Chaffee. The pressure in the cabin had risen so high that the cabin split on Chaffee's side and a roaring torrent of fire, smoke and gas surged across the lethal cabin. As the crew choked, Chaffee uttered one last cry of pain.

LEFT Vladimir Komarov was assigned to the most foolhardy space mission, *Soyuz 1*, and paid the price.

The moon at Christmas

IN 1968, NASA PLANNED ONE APOLLO 1-TYPE EARTH ORBIT MISSION, TO BE FOLLOWED BY A COMBINED APOLLO CSM AND LUNAR MODULE TEST MISSION IN EARTH ORBIT, AFTER LAUNCH ON A SATURN V.

This would be followed in 1969 by a deep Earth orbit test and a final Moon landing dress rehearsal in lunar orbit. If all went well, an American could be on the Moon by mid-1969.

Meanwhile, in Russia, the *N1* super-booster was being prepared for its maiden flight in the hope of beating America to a Moon landing, while unpiloted lunar-looping *Zond* flights were made with the aim of

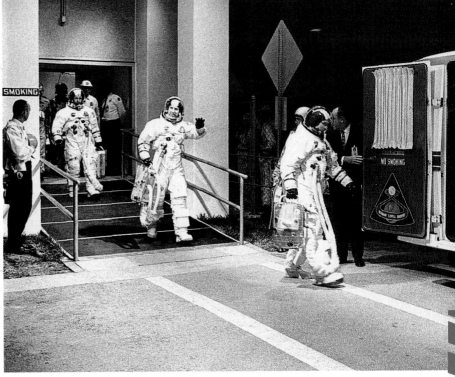

flying two cosmonauts around the Moon later in the year.

NASA was taking the Soviet threat seriously. It had spy satellite photos showing the *N1* at the Baikonur Cosmodrome and the *Zond* flights were publicized by the Soviets. As a result, even before the first manned test of the *Apollo CSM* – designated *Apollo 7* – in October 1968, the US space agency decided to fly one of the *Saturn V* tests around the Moon to beat a manned *Zond*. Although not confirmed officially until later, the *Apollo* craft would make ten orbits, rather than do just a lunar-loop and would not carry a Lunar Module. As *Apollo 7*'s commander Wally Schirra and his crewmates Donn Eisele and Walt Cunningham prepared to lift off, Russia's Alexei Leonov and a flight engineer were in training for a *Zond* lunar-loop. The Race was still on, or was it?

The belief that Russia's technology was on a par with America was still widely held in the West and this only served to perpetuate what was becoming a mythical Moon Race. Although two unmanned *Zond* 5 and 6 craft had made it back to Earth after successful lunar-loops in 1968, depressurization, the G forces and hard landings would have killed a cosmonaut crew. The previous *Zond 4* in 1968 was destroyed intentionally after an altitude control problem,

ABOVE *Apollo 8* pilot, James Lovell, centre, walks toward the launch pad with fellow astronauts Frank Borman, front, and William Anders.

LEFT Over 600 journalists covered the launch of *Apollo 7*.

BELOW "... like plaster of paris", a crater on the moon.

BELOW Bill Anders took this magnificent photo of Earth as *Apollo 8* orbited the Moon.

during re-entry, for fear it would land in "enemy hands". Another launch failed in 1968. Earlier flight tests in 1967 had not been successful. Of four launches, only one, an Earth orbit test had been successful. Unbeknown to America, it was decided that it was not safe to attempt the Leonov mission.

On 11 October 1968, *Apollo 7*, began its highly successful but rather tetchy 11-day orbital flight, clear-

APOLLO 7
Commander: Wally Schirra.
Senior Pilot: Donn Eisele.
Pilot: Walt Cunningham.
11 October 1968. Earth orbit. 10 d 20h 9m 3s

APOLLO 8
Commander: Frank Borman.
Command Module Pilot: James Lovell.
Pilot: William Anders.
21 December 1968. Lunar orbit. 6d 3h 0m 42s. Lunar orbit: 20h 11m

ing the way for the lunar orbiting *Apollo 8* aboard the first piloted *Saturn 5*. Russia then launched *Soyuz 3*, with one cosmonaut aboard to dock with the unmanned *Soyuz 2*, to perform a more modest version of the planned Komarov mission in 1967. The cosmonaut, Georgi Beregovoi, failed to dock because he approached the *Soyuz 2* upside down! But still the Western opinion was that Russia was racing to the Moon. *Apollo 8* was readied for its historic mission aboard only the third *Saturn V*. The first unpiloted test had been successful but the second in 1968 experienced problems that would have caused a crew to abort had they been aboard it. So *Apollo 8* was a pretty risky flight all round. It also became one of the most successful and historic missions in space history, making Frank Borman, Jim Lovell and Bill Anders household names. They may not have been attempting a landing but they were still the first men to the Moon. Thanks to the communications revolution taking place as a result of satellite technology, *Apollo 8* had a worldwide audience.

The *Saturn V*'s early-morning launch on 21 December 1968 was spectacular and soon the CSM was *en route* to the Moon, courtesy of the giant booster's S4B stage. The crew sent back televized programmes from *Apollo 8*, showing a fuzzy Earth from deep space. The view of the entire globe in space entranced the crew which described vividly the details they could see, especially the polar caps and weather patterns. On Christmas Eve, *Apollo 8* entered orbit, using its service propulsion system engine which fired on the far side of the Moon, out of range of the Earth. As *Apollo 8* came around the rim, Jim Lovell described the view of the Moon as "grey … like plaster of Paris".

Later, all three crewmen made an historic TV broadcast, ending it with a recitation from the first passages from Genesis in the Holy Bible. It was certainly a flight to remember, especially as a result of one photo – Earthrise. The picture of the half-Earth rising above the horizon of the barren Moon captivated millions of Earthlings, illustrating the nature of a fragile planet in the Universe, yet one that was, as Lovell described it, "a grand oasis in the vastness of space". But who took the picture? Borman says it was him, Lovell says it was him and Anders says it was him, a claim they parody comically whenever they meet or are presented in public. It was Anders! Soon after *Apollo 8* returned to Earth in triumph, NASA named the crew for *Apollo 11*, which would be the first attempt at the Moon landing. However, that depended on successful test missions by *Apollos 9* and *10*.

BELOW A flawless mission completed: US navy frogmen retrieve the *Apollo 8* capsule.

The Moon
Pathfinders

TWO PILOTED TESTS OF THE APOLLO LUNAR MODULE IN **1969** WERE VITAL TO **NASA**'S MOON LANDING PLANS. ONE TEST WAS TO TAKE PLACE IN EARTH ORBIT, THE OTHER WAS TO TAKE PLACE IN ORBIT AROUND THE MOON.

The missions were called *Apollo 9* and *10*. The first was readied for a late February launch in 1969. Meanwhile, over at Baikonur, the Soviets' last realistic chance of beating the USA to the Moon or at least stealing the thunder a little was awaiting – the first flight of the giant *N1* and dummy spacecraft simulating the *L3* Moon landing spacecraft.

On 26 February 1969, the mighty booster rose from its launch pad, beneath the thrust of 30 first-stage engines. Sixty-six seconds later, an oxidizer line to one of the engines ruptured and leaking liquid oxygen caught fire. At T+70 seconds, the engines were shut down and the launch escape system fired, carrying the simulated descent module to safety from the ensuing conflagration above the skies of Baikonur's steppe land. Again, the West did not know publicly of this dramatic event, so the assumption was that there was

RIGHT *Apollo 9*'s Lunar Module prepares for its Earth orbit test flight.

still a race to the Moon, especially as in January two manned *Soyuz* craft had accomplished what Komarov and the *Soyuz 2* were to have done in 1967 – docking and transferring crews. The fact that a crew transfer was also on schedule for *Apollo 9* may have been one reason for flying the missions. Cosmonaut Vladimir Shatalov was launched first aboard *Soyuz 4* on 14 January; the following day *Soyuz 5* roared aloft, carrying Boris Volkynov and the two spacewalkers, Alexei Yeliseyev and Yevgeni Khrunov, who were to have performed the spacewalk to *Soyuz 1* from *Soyuz 2*. *Soyuz 5* docked with the *Soyuz 4* target and the *Soyuz 5* duo walked over to *Soyuz 4* delivering some symbolic mail. That this was a demonstration of a transfer required for the Soviet Moon flight was lost on the West.

APOLLO 9

Apollo 9 got airborne on 3 March, after colds had grounded the crew. Its *Saturn 5* placed it in Earth orbit

53

BELOW Rusty Schweickart tests the Moon suit during the *Apollo 9* mission.

successfully. What followed was a brilliant technological mission which may have been seen as rather lacklustre compared with the emotion of *Apollo 8*. Everything required of a Moon landing was tested, including two astronauts, Jim McDivitt and Rusty Schweickart, test flying the lunar module, nicknamed *Spider* for the purposes of clarifying the origin of communications from two manned spacecraft. The CSM was called *Gumdrop*. The independent flight of the lunar module, including staging and the docking of the ascent stage of *Spider* with *Gumdrop*, went smoothly. Schweickart also tested the Moon spacesuit during a short EVA standing on the porch of the lunar module.

During *Apollo 9*, another interesting test was performed that did not seem that significant until a later mission. The lunar module descent stage was fired briefly while it was docked with the CSM, which may have been required in a contingency on a future mission. After *Apollo 9*, many suggested that *Apollo 10* be given a chance to make the Moon landing but the weight of the lunar module assigned to the mission was too heavy to support the fuel requirements. Although a delay of *Apollo 10* was considered, so that it could use the lighter lunar module assigned to *Apollo 11*, it was rejected in favour of a full dress rehearsal of a Moon landing in lunar orbit – everything but the landing. So, Tom Stafford and Gene Cernan did not become the first men on the Moon.

If *Apollo 10* failed for any reason, *Apollo 11*'s crew would not.

APOLLO 10

On 18 May 1969, the fifth *Saturn 5* rose ponderously from Pad 39B – the only *Apollo/Saturn 5* to leave from this pad rather than 39A – and routinely entered orbit. After a trans–lunar injection burn by the S4B, *Apollo 10* was on its way. The crew sent back jocular TV shows from the linked command and lunar modules, aptly named *Charlie Brown* and *Snoopy* respectively after the *Peanuts* cartoon characters. Then the serious business began.

Lunar orbit operations involved a simulation of the Moon landing, with the crew coming to within 14.5 km (9 miles) of the surface over a point close to where the *Apollo 11* mission was targeted in the Sea of Tranquillity. Stafford and Cernan left command module pilot John Young alone in the mother ship, coincidentally the first person to fly solo around the Moon, and set the *Snoopy* free, firing its descent engine for the first time to set up the orbital approach and then for a second to simulate the powered descent burn.

The high-spirited crew reported from *Snoopy* as they headed towards the surface that there seemed to be many, very large boulders in the target area. Just before firing the ascent engine to simulate the launch from the Moon, the spacecraft went out of control briefly because a switch had been left in the wrong position in the cockpit. It caused alarm on the ground as well as in space, particularly from Cernan who let out a fairly mild swear word for which he was reprimanded. The communications were being heard all over the world live on TV. *Snoopy*'s ensuing docking with *Charlie Brown* was successful and the mission headed home in triumph, clocking up a world record – the crew became the fastest travellers, entering the Earth's atmosphere at 39,887 kmph (24,790 mph). All was ready for the big one – the Moon landing.

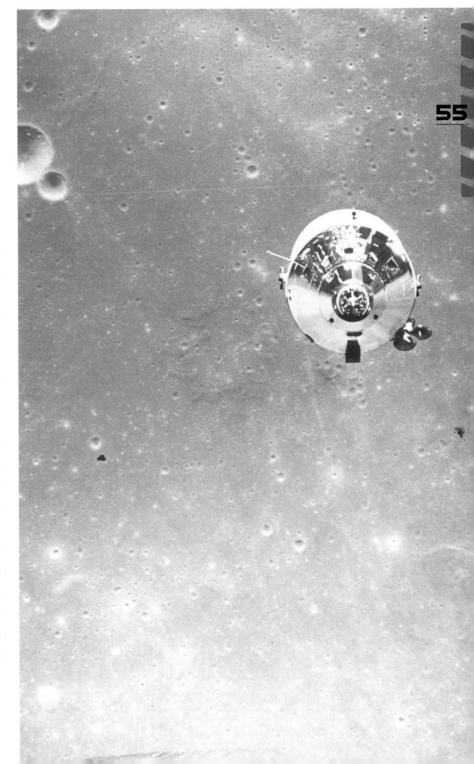

FLIGHT LOG

APOLLO 9
Commander: James McDivitt.
Command Module Pilot: David Scott.
Lunar Module Pilot: Russell Schweickart.
3 March 1969.
In lunar orbit: 10d 1h 0m 54s.
Independent flight of lunar module : 6h 20m.

APOLLO 10
Commander: Thomas Stafford.
Command Module Pilot: John Young.
Lunar Module Pilot: Eugene Cernan.
18 May 1969. 8d 0h 3m 23s.
In lunar orbit: 2d 13h 31m.
Independent flight of lunar module: 8h

55

Getting ready for the
MOON LANDING

THERE WAS NO GUARANTEE THAT APOLLO 11 WOULD MAKE THE FIRST PILOTED LANDING ON THE MOON.

Neil Alden Armstrong was born in Wapakoneta, Ohio on 4 August 1930 and joined the US Navy at 21 years of age, immediately being plunged into the Korean War. He flew 78 combat missions and during one, a wing of his Panther jet clipped a wire stretched across a valley, forcing him to eject. Armstrong later became a civilian research pilot with the National Advisory Committee for Aeronautics, later to become NASA. He flew many types of aircraft, including the seven missions in the legendary X-15 rocket plane. On one flight, he reached an altitude of 64 km (40 miles). Armstrong was selected as one of the nine second generation NASA astronauts chosen in September 1962. He served as backup command pilot of Gemini 5 and command pilot of Gemini 8 in March 1966, during which he made the first space docking in history and then almost lost his life in the spin in space. Armstrong backed up Gemini 11 and eventually became backup commander of Apollo 9 before the mission switched. The rest is history.

Command Module Pilot US Air Force Lt Col Michael Collins was born in Rome, Italy on 31 October 1930 and joined the Air Force, eventually becoming a test pilot at Edwards Air Force Base, California, where he was serving when selected to join the third group of NASA astronauts in October 1983. Collins was backup pilot for Gemini 7 and prime pilot of Gemini 10 during which he became the first person to make bodily contact with another spaceship when he spacewalked over to the abandoned Agena target vehicle used for Gemini 8. He served on some initial Apollo crews before the Apollo 1 fire and eventually found himself assigned to Apollo 8, only failing to make the flight when he injured his back. While Armstrong was a rather shy and reticent person, Collins was quite the opposite.

The third man was rather an intense and serious

That depended to a large extent on the success of *Apollo 9* and *10* but even then, the chance of total success on *Apollo 11* was regarded as 50/50 by the commander of the mission himself – Neil Armstrong. He, Mike Collins and Buzz Aldrin were not chosen specifically because they were the best astronauts but simply because they were next in the space queue. Had *Apollo 8* not made the epic lunar orbiting flight instead of a Lunar Module test in deep Earth orbit, Armstrong would have commanded *Apollo 12* and another astronaut, Pete Conrad would have been first man on the Moon as commander of *Apollo 11*. NASA selected crews in rotation, so that a crew which was backup on one flight would be prime on the third flight after. Armstrong and Aldrin were originally backups on *Apollo 9* but when it was decided to make the lunar orbit mission in 1968, the *Apollo 8* and *9* crews switched. Collins joined *Apollo 11* after he had recovered from back surgery which had taken him off the *Apollo 8* crew.

THE CREW

The *Apollo 11* trio was eminently capable of making the difficult and challenging landing mission. Each had flown one *Gemini* mission. *Apollo 11* Commander

that eventually became backup to *Apollo 8*. None of the crew would describe themselves as friends, just working colleagues with a good relationship and the dedication to complete a mission.

While Collins spent much of his time in simulating the flying of the Command Module, Armstrong and Aldrin devoted many hours practicing landings in the Lunar Module simulator. Hour after hour different problems were thrown at them as they approached "the Moon", forcing emergency aborts and even crashes, honing their skills and the skills of the flight controllers that would support the mission. If that was not enough, Armstrong also flew a strange-looking vehicle called the Lunar Landing Training

ABOVE Neil Armstrong training in the Lunar Module simulator.

individual. Air Force Colonel Edwin Eugene Aldrin was born in Monclair, New Jersey on 20 January 1930. His family nickname, Buzz, stuck. Aldrin, who served in the Korean War, held a doctorate in astronautics with a thesis on orbital rendezvous. He joined the third group of NASA astronauts in October 1963 but was surprisingly left out of the *Gemini* rendezvous and docking missions until the death of a colleague enabled him to get a seat aboard *Gemini 12* – the final *Gemini* mission. Aldrin made a record breaking spacewalk lasting over two hours from *Gemini 12* in November 1966. He joined Armstrong on the crew

Vehicle and during one flight had to eject, landing safely by parachute but only just missing the flaming crashed debris of the craft.

The crew also trained together for the phases of the mission when they flew as a group and Armstrong and Aldrin also "walked" on the Moon during well-rehearsed procedures for the two hours that they would hopefully be spending on the surface. The question of who would be first to step out on to the Moon was being aired by the press. Early checklists had the Lunar Module Pilot going out first, as the Pilot not the Command Pilot of *Gemini*s did the EVAs. However, it became obvious that it made sense for the Commander to take the first steps, if only on the basis of seniority. There have been suggestions that Aldrin took this decision badly.

OUT OF THE RACE

By July, the excitement of the prospect of the first flight to the lunar surface was building up all over the world. On 3 July, unbeknown publicly in the West, the Soviet Union finally bowed out of the Moon Race. The second *N1* booster was launched from Baikonur and suffered a massive failure just seconds after leaving the ground. It exploded, destroying the launch pad and the adjacent *N1* launch pad. Images of the destruction were captured by a US *Corona* spy satellite. The Moon Race was over. America was racing itself.

LEFT Buzz Aldrin and Neil Armstrong practice their moonwalk.

" We have LIFT OFF "

ON 16 JULY 1969, 600 MILLION

PEOPLE ALL OVER THE WORLD TUNED

IN TO THEIR TELEVISIONS TO WATCH

LIVE COVERAGE OF THE LAUNCH OF

APOLLO 11.

One million more witnessed the historic moment from the nearby beaches and causeways of the Kennedy Space Centre. This was to be one of the most significant events in human history and nobody wanted to miss it. Humans were about to set off for the first time on a journey to walk on another world in space. At 4:15 in the morning Deke Slayton, the flight crew operations director, woke Neil Armstrong, Mike Collins and Buzz Aldrin in their humble quarters in the KSC operations building on Merrit Island. It was time.

Like the countdown for their *Saturn V* rocket, sitting on the launch pad about 8 km (5 miles) away, everything was scheduled for the astronauts by the minute. Preflight medical, a 23-minute breakfast of orange juice, steak, scrambled eggs, toast and coffee, then out with the toothbrush. The crew moved to the suiting-up room and donned their custom-made spacewear. 6:26 a.m. and the crew waddled out of the operations building and into a waiting transfer van,

RIGHT The Apollo 11 crew entered the Command Module from a "white room" at the end o f the mobile service structure.

an *Apollo 11* eagle stuck to the door. The three astronauts waved at friends, fellow NASA workers and at newsmen – and photographers, taking that ominous "last photo opportunity". The van made a 24-minute journey to the Pad 39A along some of the narrow lanes that criss-crossed the tropical vegetation of the KSC. The crew got out and walked to an elevator, which rose to about 98 m (320 ft) taking them to the walkway that led across from the gantry tower to the "white room". Here they would enter the *Apollo 11* command module at the tip of the slender white giant of the *Saturn V*. It was 6:51 a.m. and in the light of sunrise, the crew could see the nearby beaches and causeways strewn with cars, vans and people, yet nearby for safety reasons the pad was almost deserted, except for a few technicians waiting to insert them into the cramped cabin. The *Saturn V* appeared to be alive. Full of propellants, it seemed to hiss and moan.

After a few cheerful farewells and traditional last minute jokes, the crew entered the capsule, first Armstrong, squeezing himself into the left hand seat,

BELOW Neil Armstrong suiting up prior to the mission.

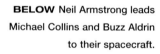

followed by Collins on the right and finally Aldrin being eased into the centre seat while backup crewman Fred Haise assisted from inside the command module. It was 7:22 a.m. Now began the rather laborious last stages of the countdown which had begun three days earlier: checking communications, programming equipment and following many procedures, to prepare the command module for launch. The hatch was closed at 7:52 a.m. Haise and the technicians had departed; the *Apollo 11* crew was alone. This was it. The final 100 minutes on the Earth.

Step-by-step, the *Saturn* was readied by the launch operations manager and his team in the control room, close to the huge Vehicle Assembly Building about 4.8 km (3 miles) away. Over two thousand of the world's press were standing at a special site close by. From here, the rocket looked about as high as a matchstick held at arms length. Five minutes to go and the access arm was retracted from the command module. The crew was wished a happy journey by the launch controller. "Thank you very much, we know it will be a good flight" said Armstrong. Fifty seconds before lift off, the *Saturn 5* 's computer was given control of the final launch sequence. Jack King, the launch commentator, read the final moments of the countdown to the listening world. At T-8.9 seconds the ignition sequence began, the *Saturn*'s five engines ignited as 36,368 litres (8,000 gallons) of water was sprayed onto the pad to cool it. The *Saturn* sat there, churning away, its engines reaching 100% thrust, spewing out flames and huge clouds either side of the pad.

BELOW Neil Armstrong leads Michael Collins and Buzz Aldrin to their spacecraft.

The lift off seemed endless.

Finally, at zero, 9:32 a.m., the monster booster was freed to fly to the Moon but it seemed reluctant to go. It rose ponderously from the pad, very slowly building up speed. "Lift off! We have lift off, 32 minutes after the hour. Lift off on *Apollo 11*. Tower clear," King reported to the world. Very slowly the *Saturn* programmed a slight yaw manoeuvre, then rolled towards its trajectory across the Atlantic Ocean.

Three miles away at the press site, it was rising in serene silence, beneath the tumultuous cloud and fire. Then suddenly, it came. The noise. An Earth-shattering roar and tremble, like ten jets on afterburner. People shouted, "go! go!", wishing the rocket into the sky. Others cried with emotion. Eventually, the thunder became a gentle murmur in the sky and the "lift off" experience was over. But it was just beginning for the three men of *Apollo 11*. Armstrong stuck to brief acknowledgements to Houston control's status reports as the rocket stages worked perfectly and the *S4B* stage and its attached *Apollo 11* stack entered orbit. "Shutdown", said Armstrong. The *S4B* later lit up again and propelled *Apollo* away from the Earth and as it shut down, Armstrong said, "that Saturn gave us a magnificent ride … we have no complaints about any of the three stages". Collins flew the *Columbia* CSM from the *S4B*, turned around and docked with the lunar module, *Eagle*, encased within it, pulling it free from the spent booster. *Apollo 11* was on its way.

THE FLIGHT TO THE MOON

The flight to the Moon was a quiet affair, the crew not knowing what lay before them, but despite this they transmitted a TV show from the interior of the command module. When some of the massive worldwide news coverage of their flight was read to the crew, mission control said that the Soviet Union had described Armstrong as the "Czar of the ship". Collins reported that "the Czar is brushing his teeth". Apollo flew on to its historic rendezvous, as the world waited.

LEFT *Saturn V* carries *Apollo 11* Moonwards from Kennedy Space Centre on the morning of July 16, 1969.

63

"The Eagle has LANDED"

AS APOLLO 11 APPROACHED THE MOON, ARMSTRONG WAS IMPRESSED BY THE SPECTACULAR SIGHT OF ALMOST THE WHOLE MOON FILLING HIS WINDOW.

"It's a view worth the price of the trip," he chortled. The craft continued its fast approach to the Moon at a speed of 1,676 m (5,500 ft) per second. The next milestone was lunar orbit insertion. Fifty minutes later, *Apollo* had "disappeared" around the far side of the Moon. The service propulsion system burst into life and burned for about six minutes reducing *Apollo 11*'s speed by about 3,218 kmph (2,000 mph), enabling it to enter a 96 km (60 mile) by 272 km (169 mile) orbit. The Moon "looks like the pictures but like the difference between watching a real football game and watching it on TV – no substitute for actually being here", said Armstrong. *Apollo 11* had entered lunar orbit safely and the excitement was building up around the world. A moment of history was just hours away – hopefully.

On 20 July, the crew had made eight lunar orbits and the time arrived for *Apollo 11* to become *Columbia* and *Eagle*, two separate craft, one which was about to perform something never attempted before – a manned landing on another world. After checking out the *Eagle* Lunar Module, Neil Armstrong and Buzz Aldrin switched the craft to internal power in preparation for its separation from *Columbia*, manned by the lone Mike Collins. The combined craft went around the far side of the moon and out of communications with ground control. A hundred hours into the mission, *Eagle* undocked. As the two craft appeared around the other side of the Moon, Armstrong reported "The *Eagle* has wings … looking good".

Further checks were made and the go ahead was made for *Eagle* to make the first of its engine firings, the descent orbit insertion burn.

Collins fired his engines and placed *Columbia* into a different orbit to await a rendezvous with the *Eagle* after the Moon landing – or before it, if a problem arose and the landing had to be aborted. Again, the Earth was cut-off from what was happening, as *Eagle* went behind the Moon. *Eagle*'s descent engine was fired and the burn lasted just 29 seconds, reducing the orbit, in preparation for the powered descent insertion (PDI) burn. *Eagle* came around the lunar rim and was given the "go for PDI" by mission control's excited, cheerful Charlie Duke, a fellow astronaut and the "cap com" for the landing. PDI began and with its engine throttling at various levels, *Eagle* flew across the Moon with its windows pointing down to the surface. Twelve or so minutes later, men could be on the Moon. No one in Mission Control expected everything to go exactly as planned but when the first alarm came, it was not one that had been anticipated or rehearsed in simulations.

At 10,668 m (35,000 ft), Aldrin in *Eagle* reported "12.02, 12.02". This was an alarm indicating that the flight computer was overloaded because *Eagle* was further down the flight path than planned and radar readings did not match what the computer was expecting and it was trying to re-programme its data. Houston recognized the problem and realized that the computer was still working as required. It was just "protesting". It gave the crew the "go" to carry on, as the engine continued to burn, eventually pitching *Eagle* towards the vertical position for the final landing phase. "Roger, we're go on that alarm", shouted Duke to the crew. *Eagle* had reduced its speed

BELOW An artist's impression of the Lunar Module separating from the Command Module.

LEFT The Lunar Module, carrying Armstrong and Aldrin, approaches the surface of the Moon.

to about 595 kmph (370 mph) when at last Armstrong could see out of his window at the terrain below, approaching fast at 40 m (132 ft) per second. Gene Kranz, the flight director asked his controllers for a final go for landing. Each controller shouted "go!". "You're for landing", said Duke to *Eagle*. Then

ABOVE The crater Dedealus taken from *Apollo 11* in Moon orbit.

suddenly, "12.01" reported *Eagle*. The computer was threatening to quit. *Eagle* continued but Armstrong noticed that it was heading for a landing in a crater the size of a football field, strewn with dangerous boulders. He took full manual control taking charge of the descent engine throttle and gimballing, plus the

craft's thrusters. Aldrin was glued to the cockpit's instrument panels and did not look out of the window, as he continued a commentary for Armstrong about the speed, descent rate and altitude, his words being heard live all over the world.

"300 ft, down 3 ½, 47 forward...altitude velocity lights …". Armstrong should have landed by now but *Eagle* was still at 23 m (75 ft) and hovering. "60 seconds!" warned Duke.

The engine's fuel was running low and the landing would have to be aborted soon. *Eagle*'s engine started to churn up lunar dust, making it difficult for Armstrong to see clearly. "Picking up some dust, 30 ft, 2 ½ down. Faint shadow. 4 forward … 4 forward … drifting to the right a little, 6 … down a half …" reported Aldrin. Mission control was silent. The tension was unbearable. Time was running out and there was nothing they could do. It was all down to Armstrong. Armstrong tried to correct the drifting as he had to land vertically or there was possibility of *Eagle* hitting a boulder and tipping over. The fuel limit was getting critical. "30s!" shouted Duke. Armstrong had to land now! Just in time, *Eagle* made contact, its engine still churning up the dust. "Contact light", said Aldrin. Speaking the first words from the Moon, he continued in a technical manner, "OK engine stop … descent engine command override, off; engine arm, off. 413 is in".

TOUCHDOWN

It was 20 July 1969, T+102 hrs 45 mins 42 seconds into the mission. There was silence. Armstrong and Aldrin slapped each other on the shoulders, smiles relieving the tension. Then Armstrong came on the line and said quietly, "Houston, Tranquillity Base here. The *Eagle* has landed".

Pandemonium broke out in Mission Control and the cigars came out. They had done it! Man was on the Moon. It was close. There was perhaps 15 seconds of fuel remaining.

LEFT The Command Module fires its boosters to ensure it goes into Lunar orbit.

67

"One Small STEP"

THE ELATED CREW OF EAGLE ATE A MEAL AND PREPARED FOR THE EPIC WALK ON THE MOON, AFTER RECEIVING PERMISSION FROM HOUSTON TO WALK ON THE SURFACE EARLIER THAN PLANNED.

OPPOSITE Buzz Aldrin stands beside the US flag on the surface of the Moon.

BELOW One small step: Aldrin's footprint on the surface of the Moon.

Armstrong and Aldrin were to have had a nap before walking but after the tension and excitement of the landing, this seemed rather superfluous. In any event, it took a long time for the crew to don their EVA suits in the cramped confines of the cockpit, the main problem being that the suit's communications antennas kept poking the ceiling, interrupting communications. The worldwide audience hearing the preparations were getting rather impatient.

Finally, the astronauts were ready and the "go" was given by Houston to depressurize *Eagle* and open the hatch. An event was about to take place during which most of the population of the world was to participate in together by TV and radio. The culmination of the Moon Race. The dreams of visionaries like Jules Verne were about to come true. Armstrong got down on his hands and knees and backed out of the hatch onto a porch, below which was a ladder leading to destiny. Guided by Aldrin, he carefully exited and came down the steps, seen on a fuzzy TV picture from the camera on the side of the lunar module. Armstrong had not decided what he was going to say when he stepped onto the surface until after he had landed because he thought the chances of making it down to the Moon at the first attempt was only about 55/50 so "why waste time" deciding.

MAGNIFICENT DESOLATION

One of the most famous quotations in history was about to be spoken. The shy and taciturn commander of *Apollo 11* jumped down onto the base of the ladder and onto a pad on one of the legs. He stepped off the pad and put his right boot onto the dusty soil – and fluffed his lines. "That's one small step for man, one giant leap for mankind", he said. He meant to say "a" man. Looking like a ghostly figure on TV, Armstrong described the soil as "fine and powdery" and later likened it to damp gunpowder. A vital act was to take a contingency sample in case he had to get back inside *Eagle* in a hurry. At least one bit of the moon would get back to the eager scientists on the Earth. Aldrin then came down the ladder to join his commander and uttered, "magnificent desolation" as he peered at the awesome sight through his tinted faceplate.

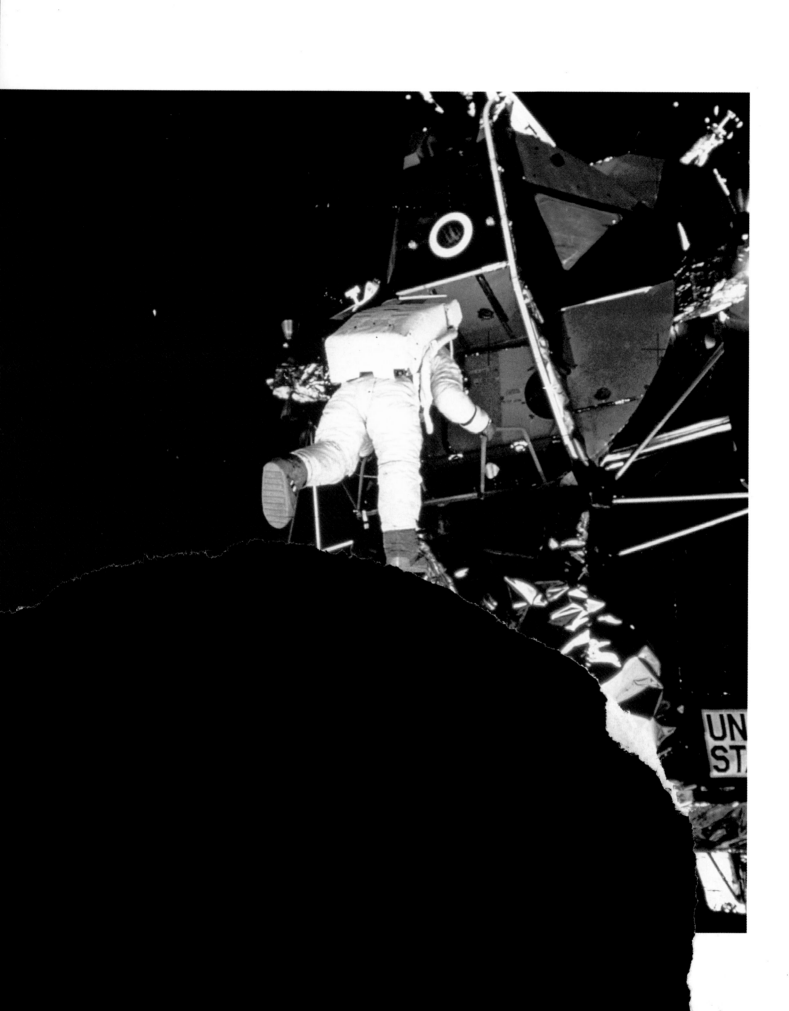

WALKING ON THE MOON

The TV camera was moved to a tripod and showed the lunar module on the surface and whenever they were in view, the two "Moonmen" waddling and making strange "bunny hops" along the surface. The crew laid out two science instruments. The Passive Seismic Experiments Package measured micrometeroid impacts on the Moon and "Moonquakes", to provide data to determine the interior structure of Earth's natural satellite. The other instrument was the Lazer Ranging Retro Reflector which reflected lazer beams from Earth back to ground stations in order to accurately measure the Moon's distance to within 15 cm (6 ins).

The Moon duo unveiled a plaque on the side of *Eagle*, which read: "Here men from the planet Earth first set foot upon the Moon, July 1969 AD. We came in peace for all mankind". President Richard Nixon

FIRST DESCRIPTION FROM THE LUNAR SURFACE

Armstrong: "The surface is fine and powdery … I can pick it up loosely with my toe. It does not adhere in fine layers like powdered charcoal to the sole and sides of my boots. I only go in...maybe an eighth of an inch, but I can see the footprints of my boots and the tread … there seems to be no difficulty in moving around as we suspected. It's even perhaps easier than the simulations at one-sixth gravity that we performed in the simulations, on the ground … the descent engine did not leave a crater of any size. There's about one foot clearance on the ground. We're essentially on a very level place here … it's quite dark in the shadow and a little hard for me to see if I have a good footing. I'll work my way over into the sunlight here without looking directly into the Sun … looking up at the LM, I'm standing directly in shadow now. Looking up at Buzz in the window and I can see everything quite clearly".

LEFT Neil Armstrong photographed Buzz Aldrin as he prepares to descend the ladder of *Eagle*.

RIGHT Aldrin was photographed by Armstrong, who is reflected in Aldrin's visor.

ABOVE, RIGHT AND FAR RIGHT Performing experiments outside the Lunar Module during the *Apollo 11* moonwalk.

joined the show when he made a live phone call to the crew, which had raised an American flag rather gingerly on a pole. "The heavens have become part of Man's world" he told the crew. Armstrong's reply was quite eloquent.

The Moonwalk was a relatively short two hours which seemed to rush by, and in no time the crew was climbing up the ladder. The show as over. The networks closed down and the feeling of anti-climax began to set in.

During the Moonwalk, Armstrong had a still camera, mounted on a bracket, attached to his chest for most of the time and took some classic photos of the Moon and of Aldrin. There is no formal photo of Armstrong standing on the Moon because Aldrin did not take one. It was not in the flight plan and the crew was so busy they did not want to waste too much time on ceremonies. However, Aldrin did have the camera for a period and he made a 360° panorama of the Moon in which Armstrong had to feature somewhere. The only picture of the commander of *Apollo 11* shows him standing by the side of *Eagle*, in shadow and with his back to the camera.

RETURN TO COLUMBIA

Attention now focused on the ascent from the Moon to join up with Collins in *Columbia*. This was perhaps the most risky part of the mission as it depended on just one engine.

The HOMECOMI

74

AMID THE EUPHORIA OF THE MOONWALK, IN THE BACK OF EVERYBODY'S MINDS WAS THE FACT THAT ARMSTRONG AND ALDRIN'S RETURN TO COLLINS IN THE COMMAND MODULE DEPENDED ON ONE OF THE VERY FEW COMPONENTS THAT DID NOT HAVE A REDUNDANT SYSTEM – THE ASCENT STAGE ENGINE.

RIGHT As the ascent stage of *Eagle* approached *Columbia*, Mike Collins took this picture of Earthrise.

If that failed, the Moonwalkers would remain on the Moon and Collins would be making a lone journey back to Earth. Armstrong and Aldrin slept fitfully in *Eagle* and were up and raring to go at T+122 hours into the mission. Their main job was to check the ascent stage engine and prepare the command for its firing. "You're clear for take-off", said Ron Evans, the capcom, sounding like an air traffic controller at an airport. Sounding like an airline pilot, Aldrin replied jokingly, "Roger, understand we're No. 1 on the runway".

Men left the Moon for the first time, as the ascent engine spurted into life and the ascent stage immediately surged skywards, its exhaust rippling the Stars and Stripes, eventually toppling it onto the swirling dusty surface that resembled the end of a busy day at the seaside – footprints and left over rubbish. Just like a *Gemini* spacecraft moving in to dock with a target rocket during the old test piloting days, *Eagle* manoeuvred to join *Columbia*. As it did so, Collins took a memorable photograph of the ascent stage and Earthrise. *Eagle* docked safely and later the two Moonmen moved into *Columbia* with their lunar spoils. The *Eagle* Lunar Module ascent stage was discarded in lunar orbit and the service propulsion system engine on the rear of *Columbia*'s Service Module burst into life, sending the three crew out of lunar orbit and back towards the Earth. As the craft came into view from the far side of the Moon, a cheerful Armstrong said, "open the LRL doors, Charlie", to capcom Charlie Duke. The LRL was the mobile lunar receiving laboratory in which the *Apollo 11* crew and their Moonrocks would remain for three weeks after landing, just in case they were carrying Moon microbes.

MISSION ACCOMPLISHED

The latter part of the *Apollo 11* mission took on a more relaxed feel; a sort of anti-climax. Kennedy's goal had been accomplished. Just the splashdown was left. The three crew members returned a TV programme *en route* to Earth with many eloquent statements about the historic flight, with Armstrong reminding everybody that it had been a team effort involving thousands of engineers, scientists and others. "The responsibility for this flight lies first with history and with the giants of sciences who have preceded this effort. Next with the American people, who have, through their will, indicated their desire. Next, to four administrations and their Congresses for implementing that will; and then to the agency and

RIGHT The full Moon as seen from *Apollo 11* during the journey home to Earth.

FLIGHT LOG

APOLLO 11

Commander: Neil Armstrong.

Command Module Pilot: Michael Collins.

Lunar Module Pilot: Buzz Aldrin.

16 July 1969. 8d 3h 18m 35s.

In lunar orbit: 2d 11h 30m.

Independent flight time by lunar module: 1d 3h 59m,
    including 21h 30m on surface.

Moonwalk time (depressurization to depressurization):
    2h 21m.

Moonrocks: 22 kg (48.5 lbs)

BELOW A spectacular Earthrise as seen from *Apollo 11*.

industry teams that built our spacecraft, the *Saturn*, the *Columbia*, the *Eagle*, and the little EMU, the spacesuit and backpack that was our small spacecraft out on the lunar surface. We would like to give our special thanks to all those Americans who built the spacecraft, who did the construction, design, tests and put their hearts and all their abilities into these crafts. To those people, tonight, we give a special thank you and to all the other people that are listening and watching tonight, God bless you. Goodnight from *Apollo 11*".

As the CSM plunged towards the Earth at 38,616 kmph (24,000 mph), the Service Module was discarded and the Command Module's thrusters manoeuvred the craft so that its heatshield was pointing at the right angle. The Command Module hit the upper reaches of the atmosphere at an altitude of about 121,920 m (400,000 ft), and the heatshield began to take the brunt of the 5,000°F (1,000°C) heat, which caused a communications blackout. The spacecraft performed a slight skip in the atmosphere, rather like a flat stone skipping on water, on a flight-path planned to reduce loads on the craft.

At 7,315 m (24,000 ft), *Columbia* was out of danger and in contact with mission control. Its two drogue chutes deployed, and a minute later, three pilot chutes followed, pulling out with them three 25.6 m (84 ft)

diameter main parachutes, which cushioned the splashdown in the Pacific Ocean, at a mission elapsed time of T+195 hr 17 min 52 sec after that tumultuous lift off. The hatch was opened and recovery frogmen from the USS Hornet threw biological isolation garments into the Command Module and shut the door again, another precaution against possible Moonbugs.

Looking like aliens from another world, the three crewmen were hoisted into a helicopter and landed on the Hornet. They made their way into the LRL and later were greeted by President Richard Nixon, who peered at the smiling faces behind the window. Nixon described the flight as the "greatest week in the history of the world since the creation" and promised that "we can reach the stars". He was to do the opposite, however. He cut the space budget, cancelling *Apollo* 18, 19 and 20. Armstrong, Aldrin and Collins left their little house, which by then had been transferred to NASA's manned spaceflight centre at Houston, Texas and re-entered the real world with a vengeance, taking off on a massive world trip which exhausted even their enthusiasm. The euphoria of *Apollo 11* was draining away quickly. The goal had been reached. *Apollo 11* was after all the third flight to the Moon. Meanwhile there had been no sign of a Russian Moonman. Russia's "sour grapes" attempt to return some samples from the Moon using an unmanned vehicle *Luna 24*, had failed during the *Apollo 11* mission. The craft crashed into the Mare Crisium. America had won the Moon Race. What else was there to do? *Apollo 12*.

77

LEFT The *Apollo 11* capsule is recovered in the Pacific Ocean.

Milk Runs and the GREAT ESCAPE

WITH THE APOLLO 11 CREW DESPATCHED ON THEIR EXHAUSTING WORLD TOUR, EFFORTS BECAME CONCENTRATED ON THE APOLLO 12 MISSION OF NOVEMBER 1969.

The Soviet Union, meanwhile, seemed to have set its piloted spaceflight sights on the development space stations in Earth orbit with further successful *Soyuz* missions. Soviet manned Moon landings seemed far off. *Apollo 12* was to demonstrate that a pinpoint landing on the Moon could be made. Its target was selected as the landing site of an unmanned pre-*Apollo* pioneer, *Surveyor 3*, which soft-landed on the Ocean of Storms in April 1967.

It promised to be quite a spectacular mission. However, after *Apollo 11*, a certain lack of interest was creeping in with the public and with only one landing achieved, *Apollo 12* was being regarded as just another milk run to the Moon. Commanded by Pete Conrad, an irrepressible little Navy test pilot, *Apollo 12* was launched into storm clouds on 12 November 1969 and not surprisingly, the *Saturn V* was struck by lightning twice within a minute of leaving the ground.

Inside the Command Module, called *Yankee Clipper*, the three astronauts were faced with something they had never prepared for in simulations. It looked like a disaster, as almost every alarm light went on in the cockpit. Thanks to mission control, order was soon restored and it seemed that *Apollo* had not been damaged after all. *Yankee Clipper*, with the Lunar Module *Intrepid* in tow, sailed to the Moon and entered orbit safely. Conrad surpassed himself and landed *Intrepid* bang on target just yards from *Surveyor*. Becoming the third man on the Moon, Conrad parodied Armstrong's words with a jokey announcement, "that may have been a small step for Neil, but its a big one for me". He hummed and whistled while he worked and soon his companion Al Bean was joining him. The TV camera however blacked out. The audiences dwindled, robbed of a chance to see the crew meeting up with "good ol' *Surveyor*", as Conrad called it, during a second Moonwalk. *Apollo 12* made it home but there was just a feeling that people had seen it all before. It looked as though *Apollo 13* would be ignored.

UNLUCKY 13

Launched on 11 April 1970 and aimed at the Fra Mauro highlands of the Moon, the mission became one of the most famous in space history because of an event that occurred 55hr 55min into the mission. It was an event that sparked off more worldwide interest and concern than the Moon landing. One of the oxygen tanks in the service module's fuel cell system exploded and as a result, the command module, *Odyssey*, slowly died, starved of electricity and oxygen and other systems. Fortunately, the crew was travelling to the Moon and the lunar module *Aquarius* was still attached. The landing was off, obviously, to the

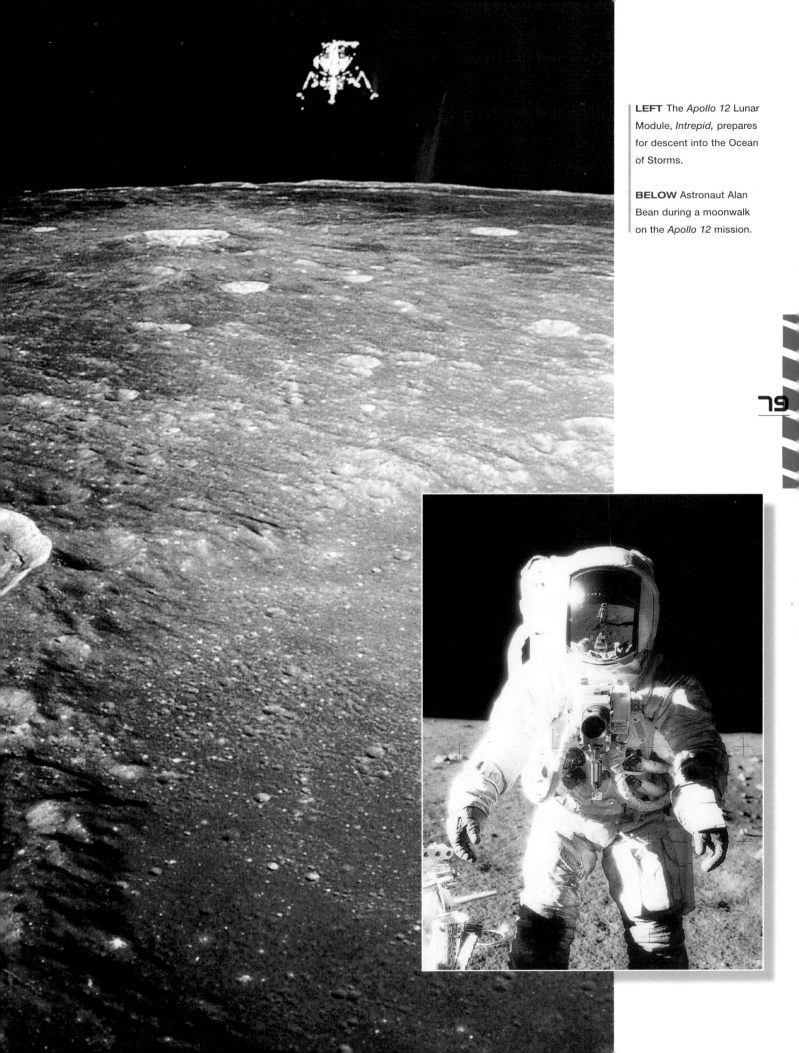

LEFT The *Apollo 12* Lunar Module, *Intrepid,* prepares for descent into the Ocean of Storms.

BELOW Astronaut Alan Bean during a moonwalk on the *Apollo 12* mission.

ABOVE The then US president, Richard Nixon, with the crew of *Apollo 13*.

disappointment of the commander, James Lovell and Lunar Module pilot, Fred Haise but *Aquarius* could be used as a lifeboat. Lovell, Haise and Jack Swigert, the Command Module Pilot who had been drafted into the crew two days before lift off to replace another astronaut who doctors feared might come down with German measles during the trip, were in desperate straits. They had to fire the *Aquarius* descent engine to place *Apollo 13* on a course which would loop the Moon and head back somewhere on the Earth. Three further burns were made to refine the trajectory, aiming for a Pacific landing – hopefully.

The flightpath was not the real worry, spacecraft consumables were. Conditions were pitiful. It was cold and carbon dioxide levels were so high that the crew would have died during the final phases of the flight had they not jury-rigged an air-scrubber devized on the ground, based on any material that the crew could get hold of in the cabin. Five days after launch, as

Apollo 13 approached Earth and the final hurdle of re-entry, it shed its lifeboat *Aquarius* and the Service Module, the damage on which could be assessed for the first time. One side of the ship had been blasted away. *Apollo 13* made it home – but only just – thanks to the efforts of thousands of engineers all over the USA, inspiring President Nixon to announce in Churchillian tones, "never have so few, owed so much to so many". The watching world breathed a sigh of relief.

Amid such drama the Soviet Union then brought its own Moon samples home in September 1970, via the unmanned *Luna 16* spacecraft, claiming that it was never in the piloted Moon landing race and it was far safer to use unpiloted spacecraft. It had also launched *Zond 7* in 1969 on the only successful unpiloted lunar-looping mission and *Zond 8* in November 1970 to end the programme. And for good measure, the same month the Soviets also launched

FLIGHT LOG

APOLLO 12
Commander: Charles "Pete" Conrad.
Command Module Pilot: Richard Gordon.
Lunar Module Pilot: Alan Bean.
12 November 1969. 10d 4h 36m 25s.
In lunar orbit: 3d 14h 56m.
Independent flight time by Lunar Module: 1d 13h 42m.
On surface: 1d 7h 31m.
Moonwalks: two lasting 7h 45m.
Moonrock cargo: 34 kg (74.7 lbs)

APOLLO 13
Commander: James Lovell.
Command Module Pilot: Jack Swigert.
Lunar Module Pilot: Fred Haise.
11 April 1970. 5d 22h 54m 41s

APOLLO 14
Commander: Alan B. Shepard.
Command Module Pilot: Stuart Roosa.
Lunar Module Pilot: Edgar Mitchell.
31 January 1971. 9d 0h 2m 57s.
In lunar orbit: 2d 18h 39m.
Independent flight time by Lunar Module: 1d 15h 45m.
On surface: 1d 9h 31m.
Moonwalks: two lasting 9h 22m.
Moonrock cargo: 44.5 kg (98 lbs)

an unpiloted lunar rover, a Heath Robinson-like machine called *Lunakhod*. Coincidentally, *Apollo 15* was to introduce the lunar roving vehicle. It all seemed a rather sad response and end to the excitement of the propaganda-based Moon Race.

NASA decided to aim *Apollo 14* at Fra Mauro, too. The mission was duly launched on 31 January 1971 and commander Alan Shepard, the *Mercury* astronaut and America's first man in space in 1961, walked on the Moon. Nobody seemed that bothered.

BELOW Alan B. Shepherd, Commander of *Apollo 14* shades his eyes from the brilliant sun during his EVA.

The End of the DREAM

THE FLIGHT OF APOLLO 15 SEEMED TO HAVE EVERYTHING: A SPECTACULAR LANDING SITE, THE FIRST LUNAR ROVING VEHICLE (LRV), AND LOTS OF SCIENCE.

Commanded by Air Force Lt Col David Scott, *Apollo 15* made a spectacular lift off from the Kennedy Space Centre on 26 July 1971 and headed in the now usual fashion *en route* to lunar orbit, hardly mentioned in the press. Unlike previous Lunar Modules which made gradually sloping descents to their targets, Scott, piloting his craft called *Falcon*, had to fly over mountains and come almost straight down into the Hadley Plain, close to St George's crater etched into the side of a

RIGHT Jim Irwin salutes to the camera at Hadley Base during *Apollo 15* with the first Lunar Rover.

FAR RIGHT Lift-off for *Apollo 15*, 26 July 1971.

mountain close to the famous Hadley Rille.

The colour TV views were spectacular as Scott and his companion James Irwin rode the first LRV, drilled the soil, picked up rocks and samples, laid out instruments and described in detail the lunar surface and rocks. They made three Moonwalks and parked the rover far enough from the Lunar Module to be able to show TV audiences the lift off of the *Falcon* ascent stage in a shower of multicoloured sparks. *Falcon* docked with the *Endeavour* command module piloted by Al Worden, who made a unique spacewalk *en route* to the Earth. All in all it was a spectacular mission.

Apollo 16 and *17* remained and would be the end of Man's first and perhaps last foray to the Moon. NASA already had its eyes on a space station called *Skylab*, which would use left over parts from *Apollo*, including a *Saturn 5* and further down the line, to a even larger multi-modular space station to be serviced by a space shuttle transport vehicle. More Moon expeditions were not on the agenda. So, it was in this climate that *Apollo 16* lifted off on 16 April 1972 for the Descartes Plains almost in the centre of the Moon as seen from Earth. Despite some technical hitches, which almost caused a cancellation of the landing, the *Orion* Lunar Module made it safely to the surface. It was not a quiet landing with Charlie Duke, as the Lunar Module pilot, excitedly talking his commander John Young down to the surface and the high spirits continued during the three Moonwalks which featured a ride on the the second LRV. The Duke–Young banter was entertaining to listen to but the scientists were disappointed when Young accidently tripped over a wire and destroyed one of the major science experiments being deployed on the surface to measure the heat within the lunar crust. Ken Mattingly, who had been dropped from *Apollo 13* after the German measles scare, was flying as command module pilot and made the now-customary EVA on the way back to Earth.

The *Apollo 16* trio made it home and there was just

FLIGHT LOG

APOLLO 15

Commander: David Scott.

Command Module Pilot: Al Worden.

Lunar Module Pilot: James B. Irwin.

26 July 1971. 12d 7h 11m 53s.

In lunar orbit: 6d 1h 18m.

Independent flight time by Lunar Module: 3d 0h 56m.

On surface: 2d 18h 55m.

Moonwalks: three lasting 18h 25m.

Moonrock cargo: 78.5 kg (173 lbs)

APOLLO 16

Command: John Young.

Command Module Pilot: Ken Mattingly.

Lunar Module Pilot: Charles Duke.

16 April 1972. 11d 1hr 51m 5s.

In lunar orbit: 5d 5h 53m.

Independent flight time by Lunar Module: 3d 9h 28m.

On surface: 2d 23h 14m.

Moonwalks: three lasting 20h 14m.

Moonrock cargo: 96.6 kg (213 lbs)

APOLLO 17

Commander: Eugene Cernan.

Command Module Pilot: Ron Evans.

Lunar Module Pilot: Jack Schmitt.

7 December 1972. 12d 13h 51m 59s.

In lunar orbit: 6d 3h 48m.

Independent flight time by Lunar Module: 3d 8h 10m.

On surface: 3d 2h 59m.

Moonwalks: three lasting 22h 5m.

Moonrock cargo: 110 kg (243 lbs)

BELOW The *Apollo 16* Lunar Module and Lunar Rover at Descartes Base.

one more mission to go and that was the end of it. Budgets were being cut and morale at NASA was low. Not as low as over in the Soviet Union, though. The Soviets lost three cosmonauts returning from the first space station, called *Salyut*, in June 1971 and two *N1* boosters failed in June 1971 and November 1972, sounding the final death knell for the Soviet lunar programme despite, ironically, some successful tests of lunar orbiter and lunar lander vehicles in Earth orbit, launched on *Proton* and *Soyuz* vehicles.

As *Apollo 17* was to be the last Moon mission, NASA dropped test pilot Joe Engle from the Lunar Module pilot position and replaced him with geologist astronaut Jack Schmitt – who would have flown *Apollo 18* had there been one. Lift off occurred on 7 December 1972 and was even more spectacular than usual as it took place at night, turning the immediate area into daylight with a mighty fireworks display and cacophony of trembling sounds. Commander Gene Cernan had a tricky landing target, the Taurus Littrow valley, which he did quite smoothly with plenty of fuel

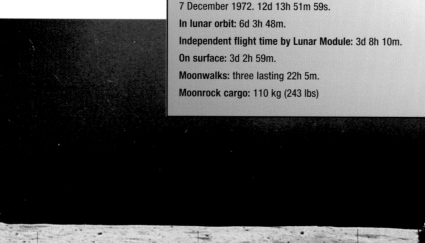

left. Cernan and Schmitt performed three Moonwalks, drove the rover for a few kilometres, picked up some orange soil and worked among boulders as large as houses, watched again by the TV camera on the rover, operated by a controller on the Earth. It was a highly spectacular and successful finale to the programme and as Cernan prepared to climb the ladder of his *Challenger* Lunar Module for the final time, to join his geologist partner, he said, "as I take these last steps from the surface, back home for some time to come, but we believe not too long into the future, I believe history will record America's challenge of today has forged man's destiny of tomorrow. And as we leave the Moon and Taurus Littrow, we leave as we came, and God willing, as we shall return, with peace and hope for all mankind".

ABOVE The *Apollo 17* crew pose for a pre-launch picture, August 28, 1972: l–r, Jack Schmitt, Gene Cernan and Ron Evans.

CHAPTER TWENTY ONE

APOLLO'S Legacy

86

ABOVE *Skylab*, the US space station that was launched in 1973.

RIGHT One of the many successful Space Shuttle missions blasts off from Cape Kennedy.

THERE IS LITTLE DOUBT THAT THE SOVIET UNION PLANNED TO LAND A MAN ON THE MOON BEFORE THE USA. THEIR DESIRE TO DO SO WAS A MAJOR FACTOR IN THE MOON RACE.

Its failure to do so was due mainly to its lack of technological sophistication and to political pressures – particularly from President Nikita Khruschev, before he was ousted in October 1964.

Khruschev and later leaders forced space teams to perform "space spectaculars", diverting them from a step-by-step approach towards their goal. By 1969, America was on its own and there was no Race to the Moon. *Apollo 11* was mainly a political goal and once achieved the space program lost momentum. Things

may have been different if a goldmine was found there, but to be honest, we did not learn much more about the Moon than before. *Apollo* was the greatest technological triumph in history. However, the enormous cost of the project drained money from other space programmes and its high profile focused public attention on this fact.

The expense was criticized mainly because the public did not feel they were getting anything out of the Moon programme. In fact, the American public spent more on their pets than on space. The irony was that while the public criticized the programme for its expense, they were watching astronauts walking on the Moon via a fleet of communications satellites, which were demonstrating the direct value of space technology to the man-in-the-street. These applications, which also included weather satellites, were overshadowed by *Apollo*. The war in Vietnam which was killing thousands of Americans, preyed on the nation's conscience. Photos of a soldier ducking Viet Cong bullets in a muddy trench, while listening to the *Apollo 11* Moonwalk seemed incongruous.

Wernher von Braun, who developed the *Saturn V* rocket, likened the *Apollo* Moonwalk to the first animals coming out of the water in the early days of Earth's history. It may have been seen like that before the Moon Race but there just did not seem to be that impression when it actually took place. Because *Apollo* was such a focused goal, there seemed to be no time for long term planning of how to follow it up and when budgets were cut no money was left to launch grandiose new ventures. *Apollo* hardware was utilized to develop *Skylab*, which was launched in May 1973. Three teams of three astronauts inhabited the station for 28, 59 and 84 days in 1973–74 before it was abandoned. Free from the Moon Race, the USSR developed its own "*Skylab*", called *Salyut*, several of which were launched between 1971 and 1982, providing a home for cosmonauts who stayed as long as 211 days.

THE SHUTTLE AND MIR

Between 1974 and 1981 no American astronauts flew in space, apart from a three person crew who flew an *Apollo* to dock with a Soviet *Soyuz* craft in 1975 in a highly political demonstration of détente between these two former arch rivals. NASA's plans for follow-on programmes after *Skylab* included a huge space station to be serviced by a space shuttle. NASA did not get any money for the station but it was granted some for the shuttle. The shuttle had nowhere to go, so it was turned into a versatile spaceship that would offer a variety of services to many users. The shuttle was given a budget that was about a fifth that spent on *Apollo* but was a vehicle five times more complicated. The resulting shuttle was not the fully reusable space taxi making 50 trips a year. It was an engineering compromise.

By June 1998, it had flown 90 missions in 17 years.

it was on the verge of cancellation. Then, in supreme irony, it was saved from the axe by Russia. After the collapse of the Soviet Union, any semblance of a Cold War had ended.

The *Mir 1* space station has been a tremendous achievement. It has enabled cosmonauts to stay in Earth orbit for as long as 430 days. These long duration flights were aimed at gaining experience for possible manned flights to Mars. But Russia was starved for cash. It was hoping to build and launch a *Mir 2* space station, to fly a reusable manned space shuttle, called *Buran*, and launch it on a giant booster called *Energia*. *Buran* and *Energia* were cancelled. Russia needed to have a space station, so it proposed a joint one with NASA. The US space agency needed to keep its station going, so agreed to an International Space Station with Russia as a major partner. Due to budget problems in both countries, the ISS is still struggling and may eventually start to be built in Earth orbit in November 1998, six years later than planned.

Apollo was a unique programme that resulted from a unique combination of factors that influenced its inception – the emergence of rocket and satellite technology, the Cold War and a bold political vision by President Kennedy. Unless that kind of combination occurs again, there will never be a programme like *Apollo*. But the irony of *Apollo*'s legacy is that in going out to the Moon, with dreams of flights to the planets and stars, the missions brought back to Earth an image that was far more emotional and beautiful than Moon rocks – the planet Earth. It was in going forth to the Moon that we were able to look back at our Earth and see it as it appears in space: a tiny, beautiful, fragile planet with an extraordinary range of life and vegetation, suspended in the Universe. Compared with the barren Moon and sterile planets, Earth is a unique place. We may go back to the Moon one day and continue to explore the planets with unpiloted craft, but somehow, there's no place like home.

LEFT Assembly of the International Space Station should be complete in 2004.

One shuttle, called *Challenger* was lost in an explosion in 1986 along with seven astronauts. The shuttle is the only piloted space vehicle America possesses. NASA eventually got the go ahead to develop a space station in 1984. It was to involve international partners and would rival the Soviet *Salyut*s and the later *Mir 1* space station, the assembly of which began in 1986. American space budgets were cut; the station was re-designed over and over again until about 1994;

CHAPTER TWENTY TWO

Where are they now?

TWELVE MEN WALKED ON THE MOON BUT NEIL ARMSTRONG IS THE ONLY NAME PEOPLE RECALL, AFTER ALL HE WAS THE FIRST.

RIGHT Neil Armstrong never sought to bask in the fame of *Apollo 11.*

Charles Lindbergh was the first person to fly across the Atlantic Ocean. Who remembers the name of the second? The Moonwalkers are largely forgotten; heroes of an extraordinary period in which Man rushed to the Moon and back, six times in 17 months, nearly 30 years ago. Neil Armstrong is often described as a recluse. He is not. A naturally shy and reticent person, he never sought the limelight. Neither did he think it right to bask in the glory of being first on the Moon. He was a representative of an *Apollo* team of thousands. He was a test pilot whose job it was to land a craft on the Moon for the first time. This for him was the highlight, not the Moonwalk.

Armstrong stayed on at NASA in an executive capacity after *Apollo 11* but then left to become a university professor. Armstrong has hosted a TV documentary series, appeared in advertisements and

supports many charitable organizations. He is now a company executive and lives on his farm in Ohio. Buzz Aldrin had a different *Apollo* aftermath. He was an intensely goal-oriented person and even during the time of *Apollo*, liked his bottle of scotch. These two things combined cruelly after his Moonwalk. After being the second person on the Moon (rather than the first), there just was not another goal to match. Even being appointed commander of the Air Force test pilot school did not help. Aldrin became depressed. He drank a lot and ended up having a breakdown, his own story being told in a book *Return to Earth*. It was turned into a film, which unfortunately created an image of Aldrin that proved difficult to shake off. Today, however, Aldrin is a healthy and happy space activist and one of the most visible of the Moonwalkers.

Apollo 12's Pete Conrad flew the first *Skylab* mission in 1973 and left NASA to work with McDonnell–Douglas in several executive capacities, until he formed his own space support company in California. Conrad is a comical storyteller. Recounting the amusing aspects of *Apollo*, he claims impishly that he and Al Bean performed the space "streak" when they transferred from the Lunar Module to the command Module after the Moonwalk. Al Bean flew a 59-day mission on *Skylab* and stayed with NASA to the point of being a candidate to command an early flight of the Space Shuttle. Bean resigned to become a highly successful commercial space artist, recreating *Apollo* Moonwalks in fine, and uniquely expert detail.

Alan Shepard, *Apollo 14* commander, became a millionaire businessman, played a lot of ProAm golf

ABOVE Buzz Aldrin, who found life difficult after *Apollo 11*, but is now a popular space activist.

and supported charitable organizations. Shepard was famous for his golf shots on the Moon. Using an improvised six iron, which he could use only one-handed, he managed to send one ball 61 m (200 ft) away in the one-sixth lunar gravity and the other ball landed in a ditch 12 m (40 ft) away, a fate known to many an exasperated golfer. Shepard died in June 1998. Edgar Mitchell concentrates his activities on studying and writing about noetic sciences and psychic research. Mitchell performed an extra sensory perception experiment during the *Apollo 14* flight unbeknown to NASA, who were most irritated when they found out after the event.

Apollo 15's David Scott headed NASA's Dryden

BELOW John Glenn, the first US man to orbit the Earth, is still involved with the space program.

Research Centre in California before creating his own aerospace company. He is working on the development of an orbital transfer module for the space shuttle. His Moonwalking colleague, Jim Irwin dedicated himself to preaching the Christian gospel as a worldwide evangelist. Irwin suffered a heart attack in 1973 and heart problems continued to dog him until he finally succumbed while walking in the foothills of his favourite Rocky Mountains on 8 August 1991, the twentieth anniversary of his landing back on Earth after *Apollo 15*.

Apollo 16 commander, John Young, went on to fly the first Space Shuttle mission in 1981 and the sixth in 1983, becoming the first person to make six space-flights. A professional astronaut, Young is still with NASA and actively training for a seventh mission at the age of 67. His chances of making another flight have been raised by the planned flight in October 1998 of *Mercury* hero John Glenn aboard the Shuttle at the age of 77 years. Charlie Duke found things hard to follow after *Apollo 16* and realized even after becoming quite rich, running a beer distribution company, that the only thing that made sense to him was the gospel of Jesus and he, like Irwin, spends much of his time as an evangelist.

Finally, *Apollo 17* commander, Gene Cernan, runs his own company and, like Aldrin, is very active on the public circuit, promoting the American space programme whenever he can. He is as patriotic now as he was when he walked proudly on the Moon. Jack Schmitt entered politics as the Senator of New Mexico in 1976 but lost his seat in 1982. He is active as a space consultant.

The even lesser known command module pilots, include the affable Michael Collins of *Apollo 11* who says he is 99% anonymous except when there is an *Apollo 11* anniversary and he is wheeled out with Armstrong and Aldrin. Two CMPs have died, Ron Evans (*Apollo 17*) from a heart attack in 1990 and Stuart Roosa, who died from complications with

93

ABOVE Michael Collins, who is perhaps the least known of the *Apollo 11* crew.

BELOW Jack Schmitt, the geologist-astronaut, who also had a political career, tests samples of Moon rock.

pancreatitis in 1994. Another CMP who died was Jack Swigert, who flew on the flight that failed – *Apollo 13*. Swigert succumbed to cancer in 1982. *Apollo 13*'s Jim Lovell has become famous through the successful film, *Apollo 13* which is based on his best selling co-written autobiography. Fred Haise tested the space shuttle in the atmosphere in 1977 and was due to command its third space mission. After delays, he left NASA and now works in industry supporting the Shuttle programme.

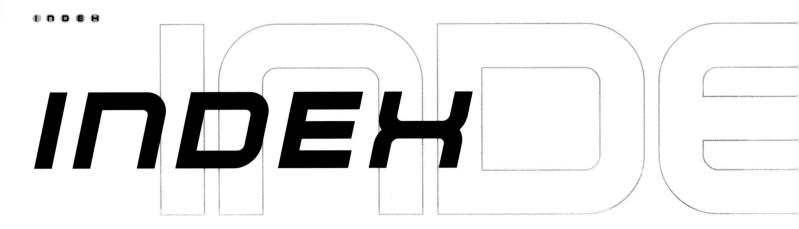

95

PICTURE ACKNOWLEDGEMENTS

The publishers would like to thank the following sources for their kind permission to reproduce the pictures in this book:

Corbis UK 12/M.Gerber 91, NASA 3, 30, Roger Ressmeyer 93l, UPI 8, 28, 73, 90, 92
Genesis Space Photo Library 6, 9, 10, 11t, 15, 17, 18, 20, 21, 22, 23, 29, 31, 32-3, 34, 35, 36, 37, 38, 39, 42, 47, 50, 52, 53, 54, 55, 64, 65, 66, 69, 75t,b, 76, 77, 79r, 80, 81, 85, 86, 87, 88, 93r/NASA 25, 26, 33, 44, 45, 48, 49, 50, 56, 57, 58, 59, 60, 61, 62, 63, 72t, ©NPO Energia 43

NASA 51, 66-7, 68, 71, 82/JPL 79l, 83, 84
Novosti Photo Library 7, 14, 16, 24, 46
Science Photo Library/NASA 4, 41, 70, 72b, Novosti 11b, 13, 19

Every effort has been made to acknowledge correctly and contact the source and/copyright holder of each picture, and Carlton Books Limited apologises for any unintentional errors or omissions which will be corrected in future editions of this book.